ENVIRONMENTAL AND WATER RESOURCES

MILESTONES IN ENGINEERING HISTORY

May 15–19, 2007
Tampa, Florida

SPONSORED BY
EWRI National History & Heritage Committee

Environmental and Water Resources Institute (EWRI)
of the American Society of Civil Engineers

EDITED BY
Jerry R. Rogers

Published by the American Society of Civil Engineers

Library of Congress Cataloging-in-Publication Data

Environmental and water resources milestones in engineering history : May 15-19, 2007, Tampa, Florida / sponsored by EWRI National History & Heritage Committee, Environmental and Water Resources Institute (EWRI) of the American Society of Civil Engineers ; edited by Jerry R. Rogers.
    p. cm.
    Includes papers of the Environmental and Water Resources Institute congress and its fourth history symposium.
    Includes bibliographical references and indexes.
    ISBN-13: 978-0-7844-0928-2
    ISBN-10: 0-7844-0928-5
    1. Water resources development--History--Congresses. 2. Water-supply--History--Congresses. 3. Environmental protection--History--Congresses. 4. Environmental engineering--History--Congresses. I. Rogers, Jerry R. II. EWRI National History & Heritage Committee.

    TC401.E575 2007
    627.09--dc22   2007012748

American Society of Civil Engineers
1801 Alexander Bell Drive
Reston, Virginia, 20191-4400

www.pubs.asce.org

Foreword

This is the fourth in a series of History and Heritage publications produced by the Environmental and Water Resources Institute (EWRI) of the American Society of Civil Engineers (ASCE). The preceding volumes were *Environmental and Water Resources History, ASCE 150th Anniversary* (2002) edited by Augustine J. Fredrich and Jerry R. Rogers, *Henry P.G. Darcy and Other Pioneers in Hydraulics* (2003) edited by Glenn O. Brown, Jurgen D. Garbrecht and Willi H. Hager, *Water Resources and Environmental History* (2004) edited by Jerry R. Rogers, Glenn O. Brown and Jurgen D. Garbrecht. These publications were sponsored by the EWRI History and Heritage Committee to provide historical engineering papers for libraries, classrooms, historians, professional societies, government agencies, firms and individuals. We encourage continued documentation of the history and heritage of the water resources and environmental professions.

Environmental and Water Resources: Milestones in Engineering History (2007) contains invited and submitted papers on a wide range of topics, many of which will be presented at the EWRI annual congress to be held in Tampa, Florida May 15-19, 2007. Unique to this volume are papers on James P. Kirkwood: environmental and civil engineer and ASCE Leader on the 200th Anniversary of his birth (1807-2007) and filtration of public water supplies in the United States. Following the past publication of histories of several university hydraulic laboratories, histories of environmental education/research are summarized at the University of Florida and the University of Texas at Austin. History and impacts of levees along the lower Rio Grande are documented. Historical small-scale water projects include those of the Mayan Indians, *acequias* of San Antonio, and the Versailles water supply at Marly. The Florida water management history project is followed by six papers on the history of experimental watershed research in the United States. As the 75th anniversary of the completion of Hoover Dam (1935-2010) approaches, a photo essay contains photos from Ansel Adams from the National Archives and Records Administration, College Park, Maryland and the Bureau of Reclamation. Two invited papers are on the hurricane of 1900 and the Galveston seawall and grade raising and recreating the July 1938 flood for computer model calibration.

The editor thanks the ASCE staff including Donna Dickert, Charlotte McNaughton, and Sheana Singletary, the EWRI staff, Tampa EWRI Congress Chairs: Paul Bizier and Karen Kabbes, reviewer and moderators: Augustine J. Fredrich, Jeffrey Bradley, William Cox, Brit Storey, and Richard Wiltshire of the Bureau of Reclamation, and Jurgen D. Garbrect for suggesting names of USDA Agricultural Research Service centers.

Jerry R. Rogers

Acknowledgments

Other contributors to this publication include the University of Houston Department of Civil and Environmental Engineering, Donna Rogers for her computer assistance which was invaluable, and Glenn O. Brown for suggesting the 2007 anniversaries which were covered in part in this publication. Not only will the 200th anniversary of the birth of James Pugh Kirkwood (born in Edinburgh- 1807) (and the second ASCE National President following fellow Scotsman James Laurie- first ASCE National President) will be celebrated in May 2007 by EWRI/ASCE, but the 250th Anniversary of the birth of Thomas Telford will be celebrated in Scotland and the U.K. Professor Roland Paxton of Heriot- Watt University- Edinburgh has coordinated the following historical events with the Institution of Civil Engineers: July 1 (Glencorse Dam Tour with Scottish Water), July 2- Telford Symposium sponsored by the Royal Society of Edinburgh and others, July 3-6 Telford Projects Tour (with Sandra Purves, including the Telford International Historic Civil Engineering Landmarks for the Caledonian Canal and Craigellachie Bridge). Civil engineers/historians of ASCE, the Canadian Society for Civil Engineering (CSCE), and the Institution of Civil Engineers will participate in the Telford250 activities. The Institution of Civil Engineers (ICE) and their Panel for Historic Civil Engineering Works have been special hosts for ASCE and CSCE civil engineers in recent years: 2003 Robert Stephenson Symposium/Projects Tour (North Wales to Chester to Newcastle to Edinburgh to the Falkirk Millennium Wheel, 2006- I.K. Brunel Symposium/Projects Tour in London/Bristol with the Newcomen Society, and 2007 Thomas Telford Symposium/Projects Tour. The Scotland civil engineers and ICE and the Panel for Historic Engineering Works seem to inform, educate, socialize with and entertain visiting civil engineers in a remarkable style. Special contributions have been by Inverness Scotsman/Canadian civil engineer Alistair MacKenzie to ASCE, ICE, and the 2005 International History Symposium in Toronto by CSCE.

As the ASCE Body of Knowledge Committee compiles its 2007 report, a Civil Engineering History and Heritage component is recommended as part of the future education of civil engineers. Special leadership was observed from Richard O. Anderson, Jeffrey Russell, water resources engineer Stuart Walesh, Melanie Lawrence, Decker Hains, Henry Petroski, hydraulic engineer Rob Ettema, and many other visionaries. The documentation, discussions, preparation, and implementation of the civil engineering Body of Knowledge may be the biggest future contribution to environmental and water resources history! Perhaps grandchildren Jackson Appel, Scott Rogers, Will Rogers and/or Jamison Appel will become engineering students of engineering history.

Contents

Photo Essay of Hoover Dam/Construction
As the 75th Anniversary in 2010 Approaches

Brit Storey, Ph.D.,[1] and Jerry R. Rogers, Ph.D., P.E., D.WRE[2]

Hoover Dam is one of the icons in civil and water resources engineering history and visited by very large numbers of people each year. The following Hoover Dam photographs were taken in the 1940s by Ansel Adams, a famous western photographer, and provided from the Bureau of Reclamation from photographs held in the WPA (Work Projects Administration) collections of the Still Pictures Branch, Archives II, National Archives and Records Administration, College Park, Maryland. Other archived Hoover Dam photos were provided by the Bureau of Reclamation.

Hoover Dam was completed in 1935 and was the tallest concrete curved-gravity (arched-gravity) dam in the world for about 30 years. Engineers came from all over the world to visit Hoover Dam and monitor the instrumentation data as the lake behind the dam was filled above the normal level to test the stresses on the concrete arch dam. In 2010, the Bureau of Reclamation will celebrate the 75th Anniversary of the completion of Hoover Dam (1935-2010). In 2002, the Bureau of Reclamation had its 100th Anniversary Celebration (1902-2002) at Hoover Dam and the University of Nevada- Las Vegas hosted a Reclamation History Symposium.

Ansel Adams was born in San Francisco in 1902, the year the U.S. Bureau of Reclamation was founded through the Reclamation Act of 1902. As a teenager, Adams visited the Pacific Exposition in San Francisco celebrating the engineering completion of the Panama Canal. For almost six decades, Ansel Adams photographed the American landscape and was recognized with awards/medals from Presidents Johnson, Carter, and Reagan. Several of his photographs of Hoover Dam are included in this photo essay along with those furnished by the Bureau of Reclamation History

---

[1]Senior Historian, U.S. Dept. of the Interior, Bureau of Reclamation, P.O. Box 25007, (Mail Code 84-5300) Denver, CO 80225-0007; phone 303-445-2918; bstorey@do.usbr.gov

[2]Department of Civil & Environmental Engineering, University of Houston, Houston, TX 77204; phone 713-743-4276; jerryrogers@houston.rr.com

Senior Historian

Archives. In 2006, the Bellagio Gallery of Fine Art covered the career of Ansel Adams though his photographs and memorabilia in an exhibit: "ANSEL ADAMS: AMERICA."

The following four Hoover Dam photographs (Figures 1-4) were taken by Ansel Adams in 1942.

Photos were provided through the Bureau of Reclamation from the WPA (Work Projects Administration) collections of the Still Pictures Branch, Archives II, National Archives and Records Administration, College Park, Maryland.

Figure 1. Downstream View of Hoover Dam with Water Flowing out the Right Outlet

Figure 2. Side View of Concrete in Hoover Dam on the Downstream Face

Figure 3. Downstream View of Hoover Dam

Figure 4. Side View of Hydroelectric Wires with Hoover Dam
in the Background

The following Hoover Dam photographs were provided from the Reclamation History archives (Figures 5 -14).

Figure 5. People in Selected Hoover Dam Photos:
Walker R. Young, Construction Engineer, U.S. Bureau of Reclamation, and Frank T. Crowe, General Superintendent, Six Companies. (Crowe was nicknamed 'Hurry-Up" by the workers. When dam construction was completed ahead of schedule, Frank Crowe received a large bonus.)

Figure 6. Construction Workers in Form Construction, with Intake
Tower in the Background

Figure 7. Dec. 1963- Floyd E. Dominy, Commissioner, 1959-1969.

Hoover Dam forms a picturesque background for Commissioner Floyd E. Dominy, who headed the Reclamation program under which the pioneer multipurpose development, and numerous other water resources projects throughout the west were constructed.

Construction of Hoover Dam Photographs

Figure 8. Night View Looking across Dam Construction from the Nevada Side

Figure 9. Downstream Face of Hoover Dam during Construction

Figure 10. June 23, 1934—Middle Portion of Downstream Face of Hoover Dam Showing Skidway Elevator and Cooling System with Header Pipes

Figure 11. Sept. 29, 1934—Nevada Intake Towers as Seen from the Upstream Cofferdam

Filling of the Lake Upstream, Water through Outlets, Completed Hoover Dam

Figure 12. Upstream Face of Hoover Dam: (Note the Rowboat in the Upstream Lake)

Figure 13. Sept. 11, 1936—Downstream Hoover Dam with Water Out of 12 Outlets

Figure 14. Jan. 14, 1946—Aerial view of Hoover Dam, which has a present installed capacity of 1,036,000 kilowatts and is the largest power plant in the world. During the war years, Boulder (Hoover Dam) produced as much as 6 billion kilowatt hours annually.

Celebrating the 200th Anniversary of the Birth (1807-2007) of James Pugh Kirkwood: Environmental/Civil Engineer and ASCE Leader

by Jerry R. Rogers,[1] Ph.D., P.E., D.WRE, F.- ASCE (Life M.)

[1] Department of Civil & Environmental Engineering, University of Houston, Houston, TX 77204 (jerryrogers@houston.rr.com (PH: 713-743-4276)

Abstract

James Pugh Kirkwood was born in Edinburgh, Scotland March 27, 1807 and died in Brooklyn April 22, 1877. In the 1830s, Kirkwood came to the U.S. and worked as an engineer on several railroads, including the expensive Starrucca Viaduct for the Erie Railroad in PA. Kirkwood became an environmental engineering specialist in municipal water works with Brooklyn, St. Louis, Cincinnati, Albany, and served for Lynn, MA in 1877. His 1869 book on water filtration practices in Europe was important to the civil and environmental engineering profession. Three sanitary (environmental) engineers: J. W. Adams, A. W. Craven (Croton Aqueduct) and Kirkwood of six civil engineers placed their names on the October 23, 1852 invitation letter to found ASCEA (and Architects) on November 5, 1852 when twelve engineers visited the Croton Aqueduct Building near Rotunda Park in NYC. Kirkwood served as an ASCEA Director from 1853-1867. In 1867 James P. Kirkwood became the second National President of ASCE. Kirkwood's inaugural ASCE Presidential address became the first ASCE paper in the first *Transactions of ASCE*.

Introduction

The suggestion for celebrating the 200th anniversary of the birth of James Pugh Kirkwood came from Glenn Brown, Oklahoma State University, who also noted 2007 will be the anniversaries of the completion of Florida's Hillsborough and West Palm Beach Canals. In the Linda Hall Library in Kansas City, Missouri, where ASCE books and publications were transferred from the NYC United Engineering Library, Glenn Brown discovered an 1869 book by James Pugh Kirkwood which he stated might be the first American publication that addressed water filtration in a systematic fashion (other than an 1835 book by Charles S. Storrow, with a small section on filters). Glenn Brown noticed that the Kirkwood book was initialed "JPK" with margin notes throughout that appeared to be notes Kirkwood made for his own use.

The 200th Anniversary of Kirkwood began with his 1807 birth in Scotland. (The year 2007 is also the anniversary of the birth of the famous Scottish Civil Engineer-Thomas Telford, the first President of the Institution of Civil Engineers (founded by eight young civil engineers. There will be Telford celebrations in Scotland July 2-8, 2007 and in other sections of the U.K. on other dates.)

Kirkwood's Early Years (1807-1830s)

James Pugh Kirkwood attended schools in Scotland (and Holland) and was apprenticed in 1821 to a local firm of civil engineers. He had attended Edinburgh College and later moved to work in Glasgow. Like his civil engineer friend- James Laurie, another Scottish engineer, Kirkwood came to the U.S. in the 1830s (Immigration Records, 1600s-1800s and A Genealogical Register....). Kirkwood began working with various U.S. railroads and authored a few railroad engineering reports.

James P. Kirkwood: Railroad/Civil Engineer (1830s-1850s) with Family Connections

James P. Kirkwood became assistant engineer on the Stonington Railroad and surveyed the route for the Long Island Railroad. Later, Kirkwood became assistant engineer for the Boston and Worcester Railroad (the initial route was built by 1833) with fellow Scotsman James Laurie working under Kirkwood (Petroski 2003). (The railroad merged with the Western Massachusetts Railroad in 1867 to become the Boston and Albany Railroad). Kirkwood worked with Julius Walker Adams, who was nephew of George W. Whistler, a famous railroad engineer and father of the famous artist: James Abbott McNeill Whistler. George W. Whistler and William McNeill were engineers involved in the building of the Carrollton Viaduct and the granite Canton Viaduct. In 1836 to provide a railroad connection to Albany to counter the Erie Canal trade, George W. Whistler had surveyed the Western Massachusetts Railroad route to connect to the Boston and Worcester Railroad. Kirkwood and Julius W. Adams observed the seven stone arch bridges in the Berkshire Mountains built by Scottish stonemason- Alexander Birnie. Kirkwood summarized an 1838-39 survey for part of the Maryland Canal (Kirkwood 1839). James P. Kirkwood married the sister of Julius Walker Adams (she passed away around 1847). When Julius W. Adams moved to the Erie and New York Railroad, Kirkwood followed and they were engineers for the Erie Railroad. [(Wisely 1974) notes that seven Presidents of the American Society of Civil Engineers worked for the Erie Railroad, including James Pugh Kirkwood (ASCE President in 1868 and Julius Walker Adams in 1875).] From their earlier experience with viaducts, Kirkwood and Adams built the Erie's famous Starrucca Viaduct, an expensive masonry construction with one of the first significant uses of concrete for the main piers (Schodek 1987). The Erie needed to reach Binghamton by the end of 1848 to fulfill its charter and Adams and Kirkwood directed 800 workers (Solomon 2002). The Starrucca Viaduct near Lanesboro, PA has 17 arches, each 50 feet wide, totaling 1040 feet long and is 26 feet wide and carries 400-ton locomotives on a viaduct

designed initially for 50-ton locomotives. With the completion of the Starrucca
Viaduct just in time for the charter, Kirkwood became Superintendent for the Erie
Railroad in 1849. For some unknown reasons, Kirkwood left the Erie Railroad to go
to the west/southwest, becoming Chief Engineer of the Pacific Railroad in St. Louis
in 1850. The Pacific Railroad was to go from the Mississippi River near St. Louis
westward over the U.S.

St. Louis Engineering, Helping ASCE Begin, and Invalid Years (1850s- 1868)

For the Pacific Railroad in 1850, James Kirkwood surveyed the route west from St.
Louis with a station now located in Kirkwood, Missouri, a city later formed around
this rail station and named after him. A Kirkwood painting hangs in the City Hall.
(There is also a James P. Kirkwood Bridge Replacement for the City of Kirkwood
designed by Horner and Shifrin Inc. to replace the former Clay Avenue Bridge.)

Although it did not limit his civil engineering work and leadership, Kirkwood was in
ill health during the last years of his life. In 1852, the year that ASCE was formed.
with fellow Scot- James Laurie and two others and early sanitary (environmental)
engineers: J.W. Adams and A. W. Craven (Croton Aqueduct), James P. Kirkwood's
name appeared on a letter: New York City October 23rd, 1852 which invited civil
engineers to attend a meeting November 5 at 9 pm in the office of the Croton
Aqueduct- Rotunda Park for the organization of a society of civil engineers and
architects (Kirkwood did not attend that meeting but it is unknown if Pacific Railroad
duties or health kept him from attending that particular meeting.) Although not
present with the founding twelve civil engineers, James P. Kirkwood was elected
Director and served as Director from 1853-1867 (Wisely 1974). At the ASCE
founding meeting in 1852, James Laurie, who was an 1848 founder of the Boston
Society of Civil Engineers, was selected as the first ASCE National President in
1853.

Brooklyn Waterworks Engineer: Environmental Publications by Kirkwood

During the 1850s and 1860s, Kirkwood was chief engineer for the Brooklyn Board of
Water Commissioners and authored numerous publications on a variety of civil and
environmental engineering topics:
(Kirkwood 1857), (Kirkwood 1857), (Kirkwood 1858), (Kirkwood 1858), (Kirkwood
1859), (Kirkwood 1862).
A summary of the Brooklyn Water Works and Sewers was printed (Kirkwood 1867).

St. Louis Water Engineer and Early Historic Water Filtration Book

In March 1865, the St. Louis Board of Water Commissioners selected James Pugh
Kirkwood as Chief Engineer. In May, Kirkwood submitted a plan for low service
pumps at the Chain of Rocks site. The City Council rejected this plan, discarded the
filter beds, and recommended the plant be located at Bissells Point. Kirkwood was
sent to Europe to study water filtration practices and which led to his environmental

engineering, historic water filtration book (Kirkwood 1869). Later in 1865, Kirkwood submitted a settling basin plan with pumps-storage reservoir-standpipe for Bissells Point that was begun in 1867. Filters were not installed at Bissells Point and only at the Chain of Rocks plant in 1915. A new Board offered Kirkwood the Chief Engineer position again, but he declined and recommended Thomas J. Whitman, brother of Walt Whitman. Kirkwood returned to NYC in 1867.

ASCE Changes with Kirkwood as the Second ASCE National President in 1867-68

Due to an unfortunate set of circumstances (some ASCE funds were lost due to poor investments in failed stocks; the ASCE President Laurie had consulting duties away from NYC; and the civil war began in the 1860s, ...), ASCE became inactive from the mid-1850s-1867 (Rogers and Ports 2002). When five shares of New York Central Railroad stock (which earned $555.25 in dividends) were found by ASCE in 1867 and President Laurie returned to NYC, Laurie called for an officers meeting on October 2 (Wisely 1974). A meeting was held on November 6 in the Society's new headquarters- the Chamber of Commerce Building, at 63 William Street. James P. Kirkwood was selected to be the second National President of ASCE. Kirkwood's Presidential Address on December 4, 1867 became the first paper published in the *Transactions of ASCE* (Kirkwood 1872). Kirkwood called for the presentation, printing, and distribution of papers at ASCE meetings (Petroski 2002). He also noted that ASCE had needed a headquarters location for many years.

During 1868, Kirkwood resigned as ASCE President and William P. McAlpine took over as the third ASCE National President on September 2, 1868 (Petroski 2002).

On December 2, 1868, the Engineers Club of St. Louis was organized, becoming an early U.S. engineering society. As 1868 National President of ASCE and since he had worked in St. Louis for years, James P. Kirkwood had a role in the founding of the St. Louis Engineers Club.

The Brooklyn Bridge, Poughkeepsi Water Filters, Albany Waterworks and A Massachusetts Report on River Pollution

In February 1869, John A. Roebling invited seven distinguished civil engineers (with $1000 per person consulting fees) to meet at the Brooklyn Gas Light Company on Fulton Street to review Roebling's plans for the Brooklyn Bridge. Among the engineers were ASCE founder Julius Adams (ASCE President in 1875), James P. Kirkwood (ASCE's second President in 1868), and William J. McAlpine (then ASCE's third President). The civil engineering board of experts discussed the bridge plans in six meetings and sanctioned Roebling's Brooklyn Bridge designs.

After 1871, Kirkwood worked on a variety of environmental engineering assignments:
1. James Kirkwood's design for a slow sand filtration system was installed in Poughkeepsie, NY.

2. He was hired in 1872 to prepare a water report for Albany, NY (Kirkwood 1872).
3. The Massachusetts State Board of Health requested Kirkwood to write a report on
the pollution of river waters (Kirkwood 1876).

James Pugh Kirkwood: 1807-1877

At the age of 70, James Pugh Kirkwood died in 1877 in Brooklyn, New York and
was buried in a cemetery there. During the later years of his productive civil
engineering career, Kirkwood was in poor health but continued as a civil engineer
and wrote many environmental reports for waterworks. Kirkwood became the second
national President of ASCE in 1867, helping revive the professional society. After
months of neglect of the grave where Kirkwood was buried in Brooklyn in 1877, a
delegation organized by the local Kirkwood, Missouri, newspaper traveled to
Brooklyn and placed a suitable headstone to honor the city's namesake. The
Kirkwood town historian wrote a summary of Kirkwood's engineering career (Cleary
1985). Born in Edinburgh, Scotland, James Pugh Kirkwood: 1807-1877 had a
remarkable career as a civil and environmental engineer and an ASCE leader.

References

A Genealogical Register of the Descendents of Several Ancient Puritans, Vol. 3,
    p. 103.

Cleary, Robert 1985. "James Pugh Kirkwood: Versatile Engineer of the Nineteenth
    Century," City of Kirkwood, MO, 41 pp.

Immigration Records (1600s-1800s): Scottish Immigrants to North America.

Kirkwood, James Pugh 1839. *Approximate Estimate of that Portion of the
    "Brookville Route" of the Maryland Canal, Surveyed during the Summer of 1838.*

Kirkwood, James Pugh 1857. *Communication from the Water Commissioners of the
    City of Brooklyn to the Mayor and Common Council, 28th August, 1857*, Hosford &
    Co.- NY.

Kirkwood, James Pugh 1857. *Communication from the Engineer to the Board of
    Water Commissioners on the Present Position of the Canal Grade, Accompanying
    Comparative Estimates of Canal and Conduit between Baisley's Pond and
    Hempstead Creek*, Brooklyn.

Kirkwood, James Pugh 1858. "Report of the Chief Engineer on the Gaugings of the
    Several Sources of Water Supply for the Brooklyn Water Works: January 1858."

Kirkwood, James Pugh 1858. "Brooklyn Water Works: Report in Relation to
    Proposals Made by Various Parties to Protect the Cast-Iron Pipes from Corrosion."

Kirkwood, James Pugh 1859. *Address of the Board of Water Commissioners to the Citizens of Brooklyn: with a Statement by the Chief Engineer*, Wm. C. Bryant & Co.- NY.

Kirkwood, James Pugh 1862. *Communication to the Board of Water Commissioners Relative to the Second Distributing Main from Ridgewood Reservoir*, L. Darbee & Son, Brooklyn.

Kirkwood, James Pugh 1867. *Brooklyn Water Works and Sewers: A Descriptive Memoir*, D. Van Nostrand, NY.

Kirkwood, James Pugh 1872. "Address of the President......December 4, 1867," *Transactions of ASCE*, Vol. I, pp. 3-6.

Kirkwood, James Pugh 1869. *Report on the Filtration of River Waters, for the Supply of Cities, as Practised in Europe, Made to the Board of Water Commissioners of the City of St. Louis*, D. Van Nostrand (Linda Hall Library-Closed Stack Books- LHL TD441 .K6 1869 ESL)

Kirkwood, James Pugh 1872. *Report of the Water Commissioners of the City of Albany: Transmitting the Reports of James P. Kirkwood*, Albany, Weed, Parsons and CO.

Kirkwood, James Pugh 1876. *A Special Report on the Pollution of River Waters*, Massachusetts State Board of Health, Reprinted in 1970, Arno, 408 pp.

Petroski, Henry 2003. "The Origins, Founding, and Early Years of the American Society of Civil Engineers: A Case Study in Failure Analysis," pp. 57-66, *American Civil Engineering History: The Pioneering Years*, Dennis, Bernie et al, ASCE.

Rogers, Jerry and M. Ports 2002. "ASCE Is Born," *CIVIL ENGINEERING*, ASCE 150th Anniversary Issue, Nov.-Dec., pp. 188-191.

Schodek, Daniel 1987. *Landmarks in American Civil Engineering*, MIT Press.

Solomon, Brian 2002. *Railway Masterpieces*, Krause Publications.

Storrow, Charles S. 1835._A Treatise on Water-Works for Conveying and Distributing Supplies of Water*, Hilliard Gray and Co.- Boston, (Linda Hall Library- Closed Stacks Books- LHL TD346 .S88 2835 ESL)

Wisely, William 1974. *The American Civil Engineer*: 1852- 1974. ASCE.

Filtration of Municipal Water Supplies in the United States

Gary S. Logsdon[1] and Thomas J. Ratzki[2]

[1]Gary Logsdon PE, 20 Springbok Drive, Fairfield, OH 45014-6616; PH(513)8601212; FAX (513)8601212; email: garylogsdonpe@earthlink.net
[2]Black & Veatch, 15450 S. Outer Forty Drive, Suite 200, Chesterfield, MO 63107; PH(636)5327940; FAX(636)5321465; email: ratzkitj@bv.com

Abstract

Developments in filtration of public drinking water supplies from after the Civil War to the present time are reviewed, based in part on information in *The Quest for Pure Water, 2nd Ed.*, published by the American Water Works Association (AWWA) in 1981. Slow sand filtration was applied during the latter 1800s. In the early 1900s after George Fuller's testing program at Louisville, the capabilities of coagulation, sedimentation, and filtration were better understood and began to be applied to municipal water supplies. Additional technical developments in the 1900s resulted in more effective and efficient water filtration and pretreatment processes. The role of James P. Kirkwood in bringing information on European water filtration practice to the United States and his influence on water treatment practice are discussed.

Introduction

Early applications of water filtration were proposed before the mode of disease transmission by microbes was understood. One of the early reasons for filtering water was the aesthetic improvement that could be attained, particularly with river waters. With the development of bacteriology in Europe in the 1870s and 1880s, and the understanding of how waterborne disease could be transmitted, the emphasis on water treatment shifted to public health protection. Thus the desire to provide safer drinking water created much of the impetus for adopting filtration. Communities that filtered their water typically had lower rates of typhoid fever during a period of several years after implementing filtration as compared to typhoid rates for several years before filtered water was provided. After slow sand filters were brought on line at Lawrence, MA; Washington, DC.; Albany, NY and Indianapolis, IN the typhoid fever death rates declined in the range of 38% to 80% (Logsdon, 1988). Following introduction of coagulation, clarification, and rapid sand filtration in the early 1900s, typhoid fever death rates at McKeesport, PA; Cincinnati, OH; Louisville, KY; Columbus, OH; and New Orleans, LA declined from 38% to 88% (Logsdon, 1988). The focus of this paper is on engineering and technology, but readers should remember that since 1900 the principal reason for water filtration has been public health protection. Advances in filtration described in this paper have also advanced public health protection.

Slow Sand Filtration and Riverbed Filtration

The successful implementation of filtration for municipal water supplies was first carried out in England and then on the European continent before being brought to the United States in the era following the Civil War (Baker, 1981). An unsuccessful attempt to filter drinking water had been made in the 1830s at Richmond, VA (Baker, pp. 127-131). Little meaningful work was done on filtration until the idea of improving the quality of drinking water in St. Louis was brought up in the 1860s.

James P. Kirkwood, a Scottish immigrant to the United States, was an engineer whose ideas for treating surface water were ahead of his time, in the context of water utility practice in the USA in the 1860s. Following a successful career as a railroad engineer in the eastern states and in the St. Louis area, he was appointed Chief Engineer of the St. Louis Water Works in 1865. A long-time employee of the St. Louis Water Works, W. B. Schworm (Schworm, undated) wrote that the Water Works had been authorized to withdraw water from the Mississippi River and convey it to the City. Kirkwood recommended that a low service pumping station be located at the Chain of Rocks on the Mississippi, and treatment by settling basins and slow sand filters. A high service pumping station and auxiliary reservoir were also in the plans. In his 1866 report Kirkwood explained that slow sand filtration had not been fully successful when used without prior sedimentation of the raw water; thus sedimentation was deemed necessary for filtration. Schworm wrote that Kirkwood recommended a filtration rate of up to 8.8 inches per hour (0.22 m/hr) and as low as 3.2 inches per hour (0.08 m/hr), which is in the range of rates presently used for slow sand filtration.

In 1866 the City Council rejected Kirkwood's plan and instead recommended that the plant, without filters, be located at Bissell Point (Schworm). This was done in 1871 but the water supplied to customers in St. Louis was still muddy. Finally, in 1894 a treatment works consisting of a river intake, low service pumping, and very large sedimentation basins was constructed at the Chain of Rocks location, thus vindicating Kirkwood's original recommendations (Schworm). To ensure that clear water would be provided for the World's Fair in St. Louis in 1904, chemical coagulation was added to improve sedimentation (Baker, p. 246). The total capacity of the six sedimentation basins was 180 million gallons (Graf, 1918), providing a theoretical retention time in excess of 24 hours. Rapid sand filters were not built at this plant until 1915, when a 160 mgd (606 ML/d) plant was placed into service (Schworm). We now know that slow sand filtration, even with plain sedimentation (sedimentation unaided by chemical coagulation), would not have been effective for treatment of river water at St. Louis. Slow sand filtration did work successfully when applied to relatively clear surface waters in New York and New England, where runoff did not carry heavy loads of silt and clay into streams and lakes and cause highly turbid surface water.

After Kirkwood's report had been prepared and sent to Council but before it was rejected, he was sent to Europe to investigate and report upon methods used there to filter water. His report (Kirkwood, 1869) was published after filtration had been rejected by Council in St. Louis. This report contained descriptions and illustrations of filtration works and filter galleries in 19 cities in England, Scotland, Ireland, France, Germany, and Italy (Baker, p. 133). Until publication of Allen Hazen's 1895 report on filtration, Kirkwood's book was the only English language text on the topic of filtration of municipal water supplies. In the Preface to the 1895 First Edition of his book (reprinted in the Third Edition), Hazen (1913) wrote about early work on water filtration and stated,

"St. Louis investigated this subject in 1866, and the engineer of the St. Louis Water Board, the late Mr. J.P. Kirkwood, made an investigation and report upon European methods of filtration which was published in 1869, and was such a model of full and accurate statement combined with clearly-drawn conclusions that, up to the present time, it has remained the only treatise upon the subject in English, notwithstanding the great advances which have been made, particularly in the last ten years, with the aid of knowledge of the bacteria and the germs of certain diseases in water."

This is a high level of praise for James Kirkwood from another engineer who made great contributions to water treatment.

James Kirkwood's work in water acquisition and treatment was not limited to St. Louis. In 1865 he recommended that sedimentation and filtration be used to treat water from the Ohio River in Cincinnati (Baker, p. 146). Later in his career he was employed as a filtration engineer at Poughkeepsie, NY (Baker, pp. 148-149) where the first municipal slow sand filter (see Fig. 1.) was constructed in the United States; at Hudson, N.Y. (Baker, p. 136); and as a consultant for water works projects involving filter galleries at Lowell and Lawrence, MA (Baker, 136). Kirkwood was referred to as "... the father of slow sand of filtration in America." by Baker (p. 148).

Figure 1. Interior View of Poughkeepsie Slow Sand Filters Undergoing Filter Maintenance, published with permission of Poughkeepsie's Water.

Another aspect of water acquisition and treatment in which Kirkwood was influential in the United States was in the use of filter galleries to acquire water of better quality than could be obtained by extraction directly from a river. His report on European practice included descriptions of five of these facilities (Baker, p. 279). No filter galleries existed in the U.S. when his extensive report on filtration was published, but evidently American engineers were impressed with his reporting on this concept, as Baker (280) reported that eleven filter galleries were built in the 1870s.

Kirkwood explained that a filter gallery could function over the long term by the filtering action as water passed down through the river bed and particles were removed; followed by flooding, when smaller particles in the river bed sediments would be suspended and carried away by the high velocity of the water, thus restoring the filtering capacity of the river bed (Baker, pp. 282-283). Kirkwood's observations about the beneficial effects of high river flow to renew the filtration capacity of the river bed have been borne out on Lake Michigan, where filtration galleries have been used in the sandy lake bottom on the Michigan coast. At least one of these has been subject to some clogging in spite of attempts to clean the lake bottom filtration gallery by flow reversal.

By the time M. N. Baker wrote Volume 1 of *The Quest for Pure Water* (the Author's Preface states that the manuscript was delivered to AWWA in 1943) many natural filter galleries had been abandoned. In closing his chapter on Filter Basins and Galleries, however, Baker wrote a short section (p. 285) entitled "Possible Revival of the Filter Gallery." Both Baker and Kirkwood were prescient in their writings about filter galleries, although the terminology and technology have changed somewhat. Now engineers use riverbank filtration and collector wells with extensive horizontal piping parallel to the river or extended out under the river to extract source water of a much-improved quality as compared to that from river intakes. A collector well having a capacity of 40 mgd has been developed by the Bureau of Public Utilities at Kansas City, KS, and the Louisville Water Company has carried out extensive research on the use of riverbank filtration as a means of acquiring higher quality water from the Ohio River.

Other engineers proposed that slow sand filtration be employed for treatment of municipal drinking water. Baker (p. 147) reported that by 1900 about twenty slow sand filter plants had been built in the USA. Many were in New York and the New England states, where low-turbidity surface waters existed. Thereafter, with the advent of rapid sand filtration, a process capable of treating muddy surface waters and requiring considerably less land area than slow sand filters for comparable production capacity, construction of slow sand filtration plants declined.

Outbreaks of waterborne giardiasis in communities that had been using supposedly pristine, low-turbidity waters with disinfection as the only treatment caused a renewal of interest in slow sand filtration in the early 1980s. Research projects funded by the U.S. Environmental Protection Agency (US EPA) showed that slow sand filters were very effective for removal of *Giardia* cysts. Subsequently slow sand filters were constructed in numerous communities in the northeastern states, the Rocky Mountain region, and the Pacific Northwest. In the 1990s, research results indicated that slow sand filters also were effective for removal of *Cryptosporidium* oocysts. Removal of both of these protozoa by slow sand filtration has substantially exceeded 99% in a number of investigations (Kohne and Logsdon, 2002).

Rapid Rate Granular Media Filtration

Following Kirkwood's publication on water filtration, the next major work on this topic was Allen Hazen's book, *The Filtration of Public Water Supplies*, published in 1895. In this era, developments in water filtration were coming at a rapid pace, and by 1913, Hazen's book had been revised twice and was in its Third Edition. That volume (Hazen, 1913) included chapters on chemical coagulation, mechanical (rapid rate) filters, removal of iron from ground waters, and water supply and disease. Hazen made many important contributions to sanitary engineering, including development of the concept effective size for filter media in sieve analysis, a theory of sedimentation that has withstood the test of time, and hydraulics tables for the loss of head in water flowing in pipes, aqueducts, and sewers (along with G.S. Williams).

Early Attempts at Rapid Sand Filtration

The American Water Works Association was founded in 1881, and in that decade the first significant attempts were being made to coagulate and filter America's surface waters at rates substantially higher than those used for slow sand filters. This work typically involved proprietary filters, equipment that might now be referred to as "package plants" or pre-engineered filters. Both pressure and gravity filters were developed. The first successful and long-term use of rapid filtration appears to have been at Somerville, NJ in the early 1880s. Accord to Baker (pp. 191-193) water filters were first installed in Somerville in 1881 or 1882. Then in 1885, four Hyatt pressure filters, accompanied by suitable coagulating apparatus, were installed. Baker reported that these filters were replaced in 1913, after having been used for 28 years.
Proprietary filters of that era employed addition of coagulant chemicals ahead of a sand filter, with little time allowed for coagulation, and no flocculation or sedimentation. Although they could treat relatively clear waters, attempts to treat muddy surface waters ended in failure. Probably the most conspicuous failure took place at New Orleans, where the filtration plant put in to service in 1893 to treat water from the Mississippi River by coagulation and filtration with no clarification was referred to as "... one of the boldest and most disastrous attempts ever made to filter the water supply of a city" by Baker (p. 205).

Coagulation, Clarification, and Filtration – Conventional Treatment

In the 1890s the Louisville Water Company sponsored trials with filtration preceded by both coagulation and sedimentation to remove a portion of the floc and sediments that would have been applied to the filter (Baker, pp. 228-236). The key to successful filtration at Louisville was the effective application of both chemical coagulation and sedimentation to remove a large fraction of the settleable solids before the water was applied to rapid sand filters. George Fuller reported that Ohio River water could be treated successfully by coagulation, clarification, and filtration at a rate of 125 mgad (2 gallons per minute per square foot {gpm/sf} or 5 m/hr). This filtration rate became the accepted rate in the United States for many years after Fuller's work had been completed; and the treatment train he employed became known as conventional treatment.

About the same time as Fuller's work at Louisville, Alan Hazen was evaluating the efficiency of chemical coagulation and filtration at Pittsburgh, PA, using Jewel and Warren proprietary filters (Hazen, p.162). He (p. 172) described the gravity filters of the 1890s as "….. a tub, with sand in the bottom and some form of drainage system." Figure 2 is an example of a wooden tub filter installation at the New Chester Water Company, constructed to filter Delaware River water.

Figure 2. Wooden Tub Filters, published with permission of Roberts Filter Group.

Hazen's filtration testing at Pittsburgh utilized laboratory analysis of plate count bacteria for assessment of filtration efficacy. Whereas turbidity measurement was still crude, bacteriology had developed to the stage where measurements of plate count bacteria could be made. Hazen (p. 179) reported 98% removal of bacteria could be attained by careful management of coagulation and rapid sand filtration. His graph of bacteria removal versus alum dosage (p. 167) showed a rapid decline in removal at alum dosages below about 8 mg/L, demonstrating the importance of effective coagulation for attaining best filter performance.

Before the municipal filtration plant was built at Louisville, George Fuller designed rectangular filters built of reinforced concrete for the East Jersey Water

Company (Baker, pp. 227-228). These went into service in 1902, and rectangular reinforced concrete filters, rather than wooden tub filters, became the preferred design concept for large municipal plants. Demonstration of successful treatment of Ohio River water provided encouragement for other cities using river sources. In the early 1900s, rapid sand filtration plants were built in Cincinnati, New Orleans, and St. Louis, among others.

With the widespread adoption of conventional treatment, water utilities and their engineers began to observe a need for improvements to coagulation and rapid sand filtration. Over time, many of these have been implemented.

Adequate cleaning of filter media is essential for effective long-term operation of granular media filters. Formation of mudballs (lumps of coagulant, dirt, and media that form in filter beds and gradually increase in size with the passage of time) can cause problems with filters. In the 1920s, John Baylis studied filter washing and found that the cleaning action during backwashing could be enhanced by water sprayed laterally into the fluidized bed by means of a fixed grid of pipes (Fulton, 1981). This was termed surface wash. Rotary sweeps were developed for surface washing in the late 1930s and generally were the mechanism of choice for surface wash, rather than grids of fixed pipes (Fulton, p. 65). About five decades later, American filtration plant designers began using the more effective European practice of providing supplemental media cleaning by air scour.

High-Rate Granular Media Filtration

The limitations of conventional treatment were brought into focus at projects involving production of very large quantities of water, such as filtration plants for Chicago and Detroit. In the early 1950s the standard rate of filtration was still considered to be 2 gpm/sf (5 m/hr) (Babbitt and Doland, 1955). Their textbook reported that the highest rates used were up to 5 gpm/sf (12 m/hr) at Chicago. Attempting to operate rapid sand filters at rates substantially greater than 2 gpm/sf resulted in trade-offs of higher headloss for conventionally-sized media or the possibility of poorer filtered water quality if coarser media were used. Stratification of filter media following backwash, with finer grains at the top of the bed where most of the floc removal and head loss increase occurred, caused this filter behavior.

To improve the capability of filters to perform at higher rates, engineers developed dual media filters consisting of a layer of larger anthracite over a layer of finer sand. Large-scale dual media installations were built and operated at Hanford, WA as a part of the Manhattan Project during World War II. This work was carried out by Roberts Filter Company (Fulton, p. 54) and Walter Conley of General Electric, and later by Neptune Microfloc. Conley (1961) reported that filters having anthracite and sand media had been used for many years before being used at Hanford, but the Hanford filters had proven capable of producing high-quality filtrate when operated at rates of 6 to 8 gpm/sf (15 to 20 m/hr). Significant aspects of the design and operation of these dual media filters included:

- Filter media consisting of 24 in (61 cm) of 0.9 mm anthracite over 6 in (15 cm) of 0.43 mm sand
- Relatively low alum feed rates

- Use of a few parts per billion of polymer for filter conditioning to strengthen floc and prevent turbidity breakthrough, (but with increased head loss)
- Systematically sampling of effluent from each filter for turbidity as measured by a light-scattering photometer

Conley reported the relative importance of process steps at the Hanford facilities to be: (1) control of alum dosage, (2) control of polyacrylamide dosage, (3) filtration, (4) flocculation, and (5) sedimentation. He stated that pilot scale results showed that when chemical dosage control and filtration were adequate, flocculation and sedimentation could be eliminated with no detrimental effect except short filter runs. He also reported that Hanford plant experience confirmed the pilot plant tests showing that pretreatment could be effective with flocculation and settling time totaling 30 to 120 minutes. The filtration studies at Hanford proved the capability of dual media filters for operation at rates up to four times greater than the so-called standard rate of 2 gpm/sf for conventional rapid sand filtration and provided incentive for filtration engineers to eliminate sedimentation basins if surface waters could be treated successfully with low doses of coagulant and filter aid chemicals. This work also led to the design of water filtration package plants that employed considerably shorter pretreatment residence times and higher filtration rates than those used in conventional treatment.

The need for filters that could operate at higher rates than 6 to 8 gpm/sf (15 to 20 m/hr) led to studies of deep bed mono-medium filters in the late 1970s and the 1980s. Filter materials larger than the typical 0.5 mm e.s. sand and 1.0 mm e.s. anthracite used in dual media filters induce lower head loss at higher filtration rates, but their filtering effectiveness for the same bed depth as a dual media bed is not as good. Therefore, deeper beds were evaluated and proven in demonstration testing by the Los Angeles Department of Water & Power (McBride, D.G. and Stolarik, G.F., 1987). The Los Angeles Aqueduct Filtration Plant, brought on line in 1987 by the Department of Water and Power, employs ozonation and direct filtration (rapid mixing, flocculation, and filtration) and was designed for rates as high as 13.5 gpm/sf (33 m/hr) based on pilot test results. The filter medium consists of 6 feet (1.8 m) of 1.5 mm e.s. anthracite. Since then other filtration plants have been built in the USA using deep bed monomedium filters, and the deep bed filter concept has been extended to dual media filters.

Improvements to Pretreatment

Effective pretreatment is a key to effective filtration, and since the early 1900s various processes involved in mixing, flocculation, and clarification have been evaluated and improved. Over time the trend for chemical mixing has been to use greater mixing intensity to thoroughly disperse coagulant chemical to the particles in the raw water and to accomplish rapid mixing in progressively shorter residence times. Flocculation, originally accomplished by flow in baffled channels, is now done by mechanical means, and a strong emphasis has been placed on attaining plug flow in the process in order to produce floc that is more nearly uniform in size, and thus possesses better settling properties. The greatest variety of treatment approaches has come about in clarification, which originally was accomplished by sedimentation in large basins with retention times of several hours or more. Since the 1930s, when equipment makers began to produce solids contact clarifiers, efforts have been made to accomplish clarification in shorter times and with processes having higher loading rates. Tube settlers were developed in the 1960s, and by providing lateral flow stability they demonstrated a workable application of Hazen's settling theory that suggested that sedimentation could be accomplished more quickly in very shallow basins. Since then clarification technologies developed in Europe or the United States have resulted in still higher loading rates, as presented in Table 1.

Table 1. Clarifier Loading (Overflow) Rates for Various Processes

Process	gpm/sf	m/hr
Conventional settling basin	0.5	1.2
Solids contact clarifier	2.5	6
Tube settler	2	5
Plate settler	3	7
Contact adsorption clarifier	10	24
Dissolved air flotation	4+	10+
Ballasted flocculation	16+	39+

Filtered Water Quality Monitoring.

Measurement of filtered water quality, a necessary task for monitoring filter performance, has evolved over the decades. Hazen (pp. 118-119) described measurement of turbidity by graduated stick into which a platinum wire with a diameter of 0.04 inch was placed. When the wire could be seen in water 1 inch below the surface, the turbidity was 1.00, and at a depth of 2 inches, the turbidity was 0.5. Reading a turbidity value less than 0.02, with the wire more than 50 inches below the water surface, was not practical. An early measurement technique that continued in use until about the 1960s was the Jackson candle turbidimeter, which was not precise for very clear filtered water. At some filtration plants, filtered water clarity was estimated using an illuminated sightwell. Typically a checkerboard pattern or some other version of a black and white tile pattern was placed in the bottom of the sightwell. This approach is similar to the Secchi Disk method used for transparency measurements in lakes. The light-scattering photometer used at Hanford is an example of early electronic instrumentation, which eliminates the human eyesight factor in measuring water turbidity. Modern turbidimeters date from about the 1960s when Hach Chemical Company introduced nephelometers for laboratory use and flow-through instruments that could produce a continuous readout of water turbidity. Continuous measurement of filtered water turbidity from each filter has been recommended in the water industry for at least two decades, and is now a US EPA regulatory requirement. This monitoring approach is especially valuable as it provides plant operators with information necessary for maintaining the highest quality of filtered water at all times.

Modern nephelometers tend to become less precise at turbidity values of 0.1 nephelometric turbidity units (ntu) or lower. This performance characteristic has led to development of a laser nephelometer which appears to be capable of ultra-low level turbidity measurement, and to use of on-line particle counters, which tend to be more sensitive than nephelometers to changes in filtrate quality at very low turbidity values.

Membrane Filtration and the Future

Filtration technology started moving in a new direction with the introduction of membrane filtration (microfiltration and ultrafiltration, or MF and UF) in the early 1990s. Both MF and UF accomplish particle removal by a straining or sieving action, and provide excellent removal of protozoa (*Giardia* and *Cryptosporidium*), bacteria, and clay particles. These filtration processes do not, however, provide for removal of dissolved contaminants unless pretreatment is managed to associate the dissolved contaminants with particulate matter. As experience is gained with pretreatment for removal of a wider range of source water contaminants in conjunction with membrane filtration, microfiltration and ultrafiltration are likely to be applied to a wider range of source waters. Filtration will continue to be a key water treatment process requiring skilled engineers for design and implementation.

References

Babbitt, H.E. and Doland, J.J. 1955. *Water Supply Engineering*, 5th Ed. McGraw-Hill, New York. 457.

Baker, M.N. 1981. Baker's 1948 book reprinted as Volume I of the 2-volume publication, *The Quest for Pure Water*, 2nd Ed. M.J. Taras, Editor. AWWA, Denver, CO.

Conley, W.R. 1961. "Experience with Anthracite-Sand Filters," *Journal AWWA*, 53(12), 1473-1478.

Fulton, G.P. 1981. "Chapter III, Filtration." Vol. II. *The Quest for Pure Water*, 2nd Ed. M. J. Taras, Editor. AWWA, Denver, CO. p. 65.

Graf, A.V. 1918. "Some Aspects of Chemical Treatment at St. Louis Water Works," *Journal AWWA*, 5, 279-287.

Hazen, A. 1913. *The Filtration of Public Water Supplies* 3rd Ed. John Wiley & Sons, New York.

Kirkwood, J.P. 1869. *Report on the Filtration of River Waters, for the Supply of Cities, as Practiced in Europe*. D. Van Nostrand, New York.

Kohne, R.W. and Logsdon, G.S. 2002. "Chapter 7: Slow Sand Filtration." *Control of Microorganisms in Drinking Water*, S. Lingireddy, Editor. ASCE, Reston,VA, 113-126.

Logsdon, G.S. 1988. "Chapter 1, Contributions of Water Filtration to Improving Drinking Water Quality," *Water Quality: A Realistic Perspective*, University of Michigan, Michigan Section AWWA, Michigan Water Pollution Control Assoc., and Michigan Dept. of Public Health, E.J.Way, Editor. 1-23.

McBride, D.G. and Stolarik, G.F. 1987. "Pilot to Full-Scale: Ozone and Deep Bed Filtration at the Los Angeles Aqueduct Filtration Plant." *Coagulation and Filtration: Pilot to Full Scale*, AWWA Seminar Proceedings No. 20017, AWWA Conference June 14, 1987. 111-147.

Schworm, W.B. undated. "A History of the St. Louis Water Works, (1764-1968)." http://www.stlwater.com/history2.html

History of the Department of Environmental Engineering Sciences, University of Florida

James P. Heaney[1]

[1]Department of Environmental Engineering Sciences, University of Florida, P.O. Box 116450, Gainesville, FL 32611-6450; PH (352) 392-0841; FAX (352) 392-3076; email: heaney@ufl.edu

Abstract

This paper presents an overview of the 60 year history of the Department of Environmental Engineering Sciences. The University of Florida, along with the University of California, Berkeley, Massachusetts Institute of Technology, Michigan State, and Rutgers were the original five graduate sanitary engineering programs in the United States. This history is divided into three periods: 1948-1966 when the program was housed in Civil Engineering; 1966-1972 when the new graduate Department of Environmental Engineering Sciences was formed; and 1972 to present with an undergraduate environmental engineering major added to the graduate program. The last part of the paper describes future goals for the Department of Environmental Engineering Sciences.

Introduction

This paper presents an overview of the 60 year history of the Department of Environmental Engineering Sciences. The University of Florida, along with the University of California, Berkeley, Massachusetts Institute of Technology, Michigan State, and Rutgers were the original five graduate sanitary engineering programs in the United States. This history is divided into three periods: 1948-1966 when the program was housed in Civil Engineering; 1966-1972 when the new graduate Department of Environmental Engineering Sciences was formed; and 1972 to present with an undergraduate environmental engineering major added to the program. The last part of the paper describes future goals for the Department of Environmental Engineering Sciences.

Sanitary Engineering: 1948-1966

The original sanitary engineering program at the University of Florida dates back at least to 1948 and was part of the Department of Civil Engineering. Some of the early prominence of the Florida program can be attributed to the efforts of two renowned faculty, A. Percy Black and Earle B. Phelps. A.P. Black was a University of Florida Professor of Chemistry and Sanitary Science for 47 years. He was a leader in the fields of water chemistry and water treatment. Dr. Black founded Black, Crow, and Eidsness Consulting Engineering firm which was later acquired by CH2M-Hill. He served as President of the American Water Works Association in 1949-1950. Earle B. Phelps was a Professor of Sanitary Science from 1944 to 1953 when he passed away at the age of 76. He is considered to be the father of the field of Public Health Engineering based on his two textbooks on this subject (Phelps 1948, and Phelps and Tiedeman 1950). Phelps also was a pioneer in water quality management. He published the results of much of his work in a book titled *Stream Sanitation* (1953) that is based on his classic work with H.W. Streeter on the Ohio River that led to the Streeter-Phelps equation for estimating dissolved oxygen concentrations in rivers and streams (Streeter and Phelps 1958). Vesilind (2006) describes Phelps' leadership in formulating the coliform standard for drinking water.

The original sanitary engineering program focused on the following areas: public health engineering, stream sanitation, paper and pulp industry wastes, water and wastewater treatment, radiological health, eutrophication, water chemistry, and air pollution. The research on industrial wastes was done in cooperation with the National Council for Air and Stream Improvement that was located on campus at the Phelps Lab. The Sanitary Engineering Program was very successful in launching a graduate program. They sought to add faculty in chemistry and biology to broaden the program. Other Civil Engineering faculty resisted. They felt that all of the faculty should have terminal degrees in Civil Engineering. In 1996, the Sanitary Engineering group broke off from Civil Engineering to form a separate Department of Bioenvironmental Engineering Sciences. The name was changed a few years later to Environmental Engineering Sciences.

Department of Environmental Engineering Sciences Graduate Program: 1966-1972

With generous support from the U.S. Public Health Service and its successors, the department grew rapidly beginning in the mid 1960s to about 16 full time faculty in 1972. These 16 faculty were organized around federal major training and research grant programs in the following areas with the number of faculty shown in parentheses: air pollution (3), water and wastewater (3), environmental biology (3), water resources (3), radiological health (2), solid waste (1), and ecology (1). The majority of the total support for these programs, including faculty salaries, came from the federal government. The government provided the initial support for several years with the intent that these programs would eventually become self-sustaining.

Professor H.T. Odum joined the faculty in 1970 as a Graduate Research Professor. Professor Odum was one of the founders of modern ecology, environmental science, ecological engineering and environmental economics. He produced 15 books, nearly 300 articles and was chairman for 75 doctoral dissertations at the University of Florida from 1970 to 2000. He and his brother, Eugene Odum, won the Crafoord Prize in 1987, equivalent to a Nobel Prize in ecology. *Environment, Power and Society* (Odum 1971) was a cornerstone text for the environmental movement of the 1970s and is still relevant today. H.T. Odum founded the Center for Wetlands in 1973. This center became an international focal point for research that demonstrated the value of wetlands and helped launch numerous efforts to protect and restore them. Professor Odum passed away in 2002.

Department of Environmental Engineering Sciences: 1972-Present

The early 1970s marked a major transition and growth of the entire environmental field. The U.S. Environmental Protection Agency was created and transformed the environmental field to one dominated more by construction grants and more aggressive regulation of environmental standards. Under EPA, the focus also shifted from primary concern with public health to environmental health. An unprecedented federal construction grants program provided the majority of the costs for constructing secondary and tertiary wastewater treatment plants for municipal wastes. The 1970s also marked the beginning of interest in nonpoint pollution.

The EES Department created an undergraduate program in 1972 in response to the need for a much larger number of environmental engineers. This program has been one of the largest of its kind since its creation. Fall undergraduate enrollments in our ABET accredited environmental engineering major since 1988 are shown in Figure 1. Interest in environmental engineering was relatively low during the 1980's but began a rapid ascent during the early 1990's due to increased interest in the field and federal funding in several areas including hazardous wastes. Enrollments increased from 92 juniors and seniors in 1988 to a high of about 350 undergraduates in 1996. Following the peak in 1966, undergraduate enrollments declined at about the same rate that they had increased reaching a low of 81 students in 2003. In Fall 2006, enrollments bounced back to 130 undergraduates. The undergraduate curriculum is designed to expose students to all aspects of environmental engineering. Accordingly, the students have few electives.

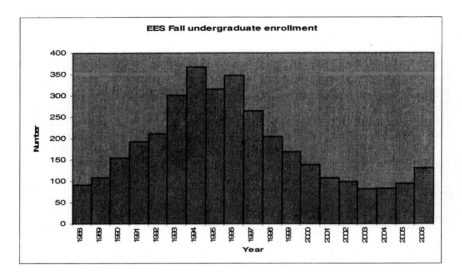

Figure 1. Undergraduate Environmental Engineering majors in the Department of Environmental Engineering Sciences from 1988 to 2006. Data for 1988 to 1995 are for juniors and seniors only.

Enrollments in the EES Master of Engineering and Master of Science ;programs have followed the same trends as the undergraduate enrollment with a rapid increase in the early 1990's followed by a steady decline up to 2003 as shown in Figure 2. During the past three years, enrollments in the Masters program have increased to about 60 students. Most of this enrollment increase is attributable to the addition of an online masters program. Enrollments in the Ph.D. program averaged about 15 students from 1988 to 2002. They have quadrupled in the last four years to 65 Ph.D. students in Fall 2006 as shown in Figure 2. This gain was stimulated by a concerted effort to increase Ph.D. enrollments.

Graduate students have wide flexibility in selecting their courses and research areas. Graduate programs are divided into seven thrust areas: Air Resources, Bioenvironmental Systems, Ecological Systems, Solid and Hazardous Waste Management, Water Resources, Water Supply and Wastewater Systems, and Environmental Nanotechnology. Specific course requirements are worked out with the graduate student and their advisor. The online masters program offers majors in two areas: Water Resources Planning and Management; and Water, Wastewater and Stormwater Engineering. Required courses for these online programs are scheduled 3 to 5 years in advance to allow students to plan their program areas.

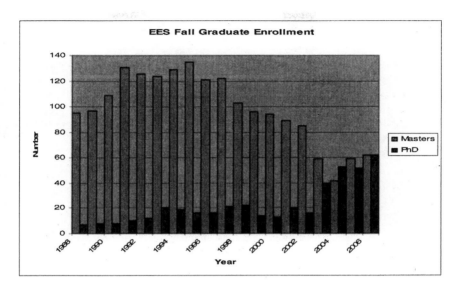

Figure 2. Graduate enrollment in the Department of Environmental Engineering Sciences from 1988 to 2006.

Sponsored research generated by the departmental faculty ranged between $1 and $1.5 million per year during the 1990's as shown in Figure 3. Concerted efforts to increase sponsored research during the past five years have increased the annual expenditures to about $3 million per year. An Environmental Systems Commercial Space Technology Center was created in 2000 with support from NASA. This center has generated a nice mix of research and commercialization of this research in collaboration with private sector organizations. The EES Department also hosts two other centers: the Center for Wetlands, and the Center for Environmental Policy.

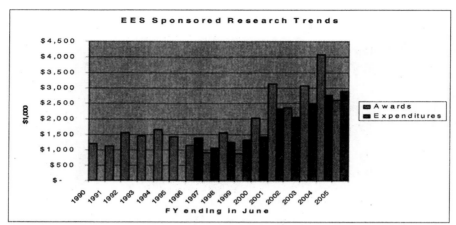

Figure 3. Trends in sponsored research at the University of Florida between 1990 and 2006.

Buildings and Field Sites

The original Sanitary Engineering program was housed in the 5,500 square foot Phelps Lab. In 1966, the newly created Department of Bioenvironmental Engineering Sciences moved into the new four story, 38,000 square foot, A.P. Black Hall that is shown in Figure 4. The most recent major expansion was the addition in 1996 of 12,600 square feet of space in the New Engineering Building located adjacent to Black Hall.

Figure 4. A.P. Black Hall at the University of Florida that was completed in 1966.

In addition to over 56,000 square feet of laboratory and office space for the Department of Environmental Engineering Sciences, an on-campus water reclamation plant and the campus watershed itself are used as living laboratories for research and teaching.

Summary and Conclusions

The Department of Environmental Engineering Sciences has a long and distinguished history of making major contributions to the profession and citizens of the State of Florida, the United States, and countries throughout the world. Detailed information about the department is available on our web page (http://www.ees.ufl.edu). The total number of 18 faculty has remained virtually constant during the past 40 years. We continue to be blessed with first-rate facilities to house our activities. The number of students has varied widely over the years as interest in the field waxes and wanes. The entire environmental field has exploded during the past 40 years. In the 1960's, our department was the only one at the University of Florida conducting research and education in environmental subjects. Now, numerous environmental programs exist across campus with over 275 faculty with interest in some aspect of the environment. Similar situations have occurred across the United States. Our goal for the future is to pursue excellence in both undergraduate and graduate education by focusing our efforts in areas where the need is greatest and we have the capacity to excel. The world is facing unprecedented environmental problems. We can help solve them.

References

Odum, H.T. 1970. *Environment, Power, and Society*. J. Wiley and Sons, NY, 331 p.

Phelps, E.B. 1948. *Public Health Engineering-A Textbook of the Principles of Environmental Sanitation. Vol. I*, J. Wiley and Sons, NY, 655 p.

Phelps, E.B. 1953. *Stream Sanitation*. J. Wiley and Sons, NY, 276 p.

Phelps, E.B. and W.D. Tiedeman. 1950. *Public Health Engineering. Vol. II*, J. Wiley and Sons, NY, 213 p.

Streeter, H.W. and E.B. Phelps. 1958. *A Study of the Pollution and Natural Purification of the Ohio River*. U.S. Dept. of Health, Education, and Welfare, Washington, D.C.

http://dspace.udel.edu:8080/dspace/bitstream/19716/1590/2/C%26EE148.pdf

Vesilind, P.A. 2006. "Expediency and the evolution of the coliform standard for drinking water". *AEESP Newsletter*, April, p. 4.

History of Environmental Engineering
at
The University of Texas at Austin

Joseph F. Malina, Jr., Ph.D., P.E., DEE, F. ASCE
C.W. Cook Professor in Environmental Engineering

Earnest F. Gloyna, D.Eng., P.E., DEE, F. ASCE
Betty Margaret Smith Chair in Environmental Health Engineering, Emeritus

Department of Civil, Architectural, and Environmental Engineering
The University of Texas at Austin (512-471 6414, http://www.utexas.edu)
Austin, TX 78712-0273

Abstract

Sanitary engineering was a distinct component of engineering when the Department of Engineering was created at The University of Texas at Austin in 1894. In 1936 the sanitary engineering option in the Civil Engineering program was accredited by the ECPD and continues today as an option in the undergraduate ABET accredited Civil Engineering program.

The graduate program in Sanitary Engineering was approved in 1947 by the Graduate School and in 1962 the name was changed to Environmental Health Engineering and the EHE program was accredited at the advanced level (M.S.E.) by ABET in 1963. The name of the graduate program was changed to Environmental and Water Resources Engineering (EWRE) in 1998. EAC/ABET accredited the MSE –EWRE program in 2005 through 2011 resulting in more than 48 years of continuous accreditation of the environmental engineering graduate program. This paper focuses on the history of the graduate program.

Introduction

Sanitary engineering education in Texas started with the establishment of The University of Texas at Austin at the end of the 19th century. The 1894-1895 Department of Engineering Catalog announced four distinct groups of courses (Civil, Sanitary, Electrical and Mining Engineering). Graduate work was offered in Hydraulic and Sanitary Engineering. The civil engineering student took some courses dealing with water supply, water filtration, sewerage systems, and sewage disposal as they are required to do in the 21st century. It has been reported that one of the 1907 graduates, identity unknown, majored in sanitary engineering. The 1903-

1904 catalog identifies specific courses required for sanitary engineering in each of the four years.

Sanitary engineering flourished as an integral part of the civil engineering program. In 1936 the Sanitary Engineering Option in the Civil Engineering program was accredited by the Engineering Council for Professional Development (ECPD). The name of the option evolved into environmental engineering and expanded beyond water supply and treatment and sewerage collection and sewage disposal to include water quality issues, environmental sampling and analysis, water and wastewater treatment plant design, indoor and outdoor air pollution control, solid waste engineering and management, and hazardous waste remediation. Environmental Engineering continues in the 21st century as an option in the undergraduate Civil Engineering program that has been accredited for more that 70 years by the Engineering Accreditation Commission of the Accreditation Board for Engineering and Technology (EAC/ABET).

Master of Science in Engineering (M.S.E.) Program

The graduate program in Sanitary Engineering initially was an option in Civil Engineering. The roots of the graduate program can be traced to Mr. Quinton B. Graves who joined the Civil Engineering faculty in the 1937 to teach hydraulics and sanitary engineering. The first master's degree in sanitary engineering as an accredited option in civil engineering was awarded to Herman Kelley Clark in 1939. However, it was not until 1947 that the graduate program in Sanitary Engineering was approved by the Graduate School at The University of Texas at Austin. It is believed that the impetus for seeking approval of the Graduate School for the sanitary engineering masters program may have originated with military veterans (GIs) who returned to The University after serving in World War II, specifically, Ben B. Ewing and Earnest Gloyna who joined the faculty as Instructors in 1947 and were promoted to Assistant Professors in 1949 after earning MSCE degrees in sanitary engineering under the tutelage of K.W. Cosens. Professor Ernest W. Steel joined the faculty in the fall of 1950. Professor Steel brought with him wide experience in teaching, writing, and professional practice. He co-authored with Mr. Victor Ehlers, State Sanitary Engineer, Texas State Department of Health, a nationally recognized sanitary engineering textbook.

Earnest F. Gloyna went on the earn a Doctor of Engineering degree at Johns Hopkins University in 1953 and returned to The University to resume his teaching and research activities in the sanitary engineering program. This event marks the beginning of the graduate program in sanitary engineering at The University of Texas at Austin as we know it today. Professor Steel retired in the fall 1960 and Joseph F. Malina, Jr. joined the faculty in the spring 1961 as an assistant professor launching his long career in the environmental engineering program at The University.

An active research program was initiated, graduate student enrollment increased, course offerings expanded to include public health engineering, water quality issues, water sampling and analysis, advanced water and wastewater treatment, air pollution control, and radiological health. The name sanitary engineering was not fully descriptive of this growing discipline. In 1962, the name formally was changed to Environmental Health Engineering (EHE) to more completely describe the growing program and to reflect the scope of the academic work, research activities, and related studies in water resources, air resources, and environmental health that are included in the program and recognize the effects of man's activities on human health.

The EHE program was accredited at the advanced level (M.S.E.) by the Engineering Accreditation Commission (EAC) of the Accreditation Board for Engineering and Technology (ABET) in 1963. The EHE graduate program has enjoyed continuous ABET accreditation ever since. The EHE graduate program expanded to incorporate water resources engineering and the name of the graduate program was changed to Environmental and Water Resources Engineering (EWRE) in 1998. EAC/ABET accreditation of the MSE – EWRE program continue and currently the program has been accredited through 2011 resulting in more than 48 years of continuous accreditation by ABET of the environmental engineering graduate program. The EWRE program is one of the oldest EAC/ABET accredited programs in the United States. The EWRE graduate program, although accredited by EAC/ABET at the advanced level, is an integral part of the Civil, Architectural and Environmental Engineering Department.

Early research facilities include the Environmental Health Engineering Research Laboratories (EHERL) at the Balcones Research Center (Pickle Research Campus) and a temporary building on the main campus. The Center for Research in Water Resources (CRWR) was established in 1963 to reflect the expanding research efforts that focused on water quality and water resources of Texas. The research facilities were consolidated in 1970 in the newly completed CRWR facilities at the Pickle Research Campus. Additional space became available to the EWRE program faculty and students in the newly completed Ernest Cockrell Jr. building on the main campus in 1975.

The ingredients for a strong engineering program were falling into place, namely, excellent students, an EAC/ABET accredited academic program, excellent facilities with well-equipped laboratories, state-of-the-art instrumentation and superb computation facilities, and a qualified faculty. A chronology of faculty who taught sanitary, environmental and water resources engineering in the graduate program in Environmental and Water Resources Engineering are listed in Table 1. Currently there are 16 full-time faculty members and one Adjunct Professor. There are five members of the National Academy of Engineering (NAE) among the active full-time, adjunct, and emeritus faculty members.

The key to the success of the graduate program at The University of Texas at Austin is the strong engineering emphasis. Graduate studies in EWRE encompass treatment process engineering and design, water resources engineering, hydraulic engineering design, air resources engineering, indoor air quality, environmental remediation and risk assessment, solid waste engineering and management, water quality of natural waters and pollutant transport modeling.

The graduate programs require no specific courses; therefore, each student tailors a program of work that satisfies personal education and professional goals. Each student is encouraged to build a program of work that helps the individual prepare to meet personal career objectives and provide information necessary to meet the scope of the thesis or departmental report. The program focuses on educating engineers who will solve environmental and water resources problems by applying fundamental principles from natural sciences, mathematics, mechanics, engineering, economics, and other underlying disciplines. Graduates are prepared to solve both current and future environmental problems. Graduate studies in EWRE encompass treatment process engineering (water, wastewater, and hazardous waste treatment), water resources engineering (hydraulics, hydrology, GIS, hydraulic engineering design), air resources engineering (air pollution sources and control, indoor air quality), environmental remediation (soil & ground water clean up, risk assessment), solid waste engineering and management and water quality (contaminants in natural waters, pollutant transport modeling). All graduates of the MSE (EWRE) program must complete a thesis, departmental engineering report or a master's report.

Admission to the MSE (EWRE) program requires an earned bachelor's degree in engineering from an EAC/ABET accredited engineering program. However, applicants with a degree in natural sciences or mathematics with outstanding undergraduate academic records may be admitted to the MSE (EWRE) program after completion of required remedial undergraduate engineering coursework.. Additional requirements are designed to ensure that graduates have knowledge of fundamental engineering principles and can meet the minimum requirements for application for the Engineering Fundamentals examination for certification as an Engineer-in-Training in the registration process as a professional engineer.

A total of 991 M.S.E. degrees in sanitary engineering, environmental health engineering and environmental and water resources engineering were earned during the period of 1937 through the summer 2006 A summary of the number of M.S.E. graduates earned in 5-year time increments is presented in Figure 1. These data indicate that on average 20 engineering degrees were earned each year during this time period. However, during 1991-2005 the average number of M.S.E. degrees was approximately 30 per year. Data collected from a survey of M.S.E. EHE and EWRE graduates indicate that 48% of all graduates responding and 64% of those graduating since 1998 were employed in consulting engineering, 11% of all responses and 7% of those graduating since 1998 were employed in industry. The strong process engineering and engineering design components of the education experience prepare the graduates for productive careers in the environmental engineering profession. Those responding who were employed by municipal, state or federal government represented 11% of all and 4% of the more recent graduates. On the average about 14% of the recent graduates went on to graduate school.

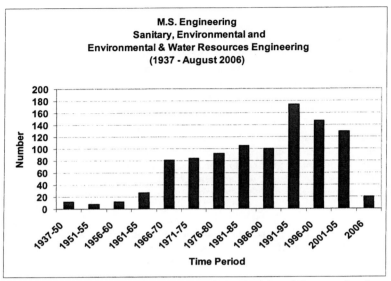

Figure 1. M.S.E. Degrees Awarded at The University of Texas at Austin
(Sanitary, Environmental & Water Resources Engineering)
(1937-August 2006)

Doctoral Program

The Ph.D. program is in Civil Engineering with emphasis in sanitary engineering, environmental engineering or water resources engineering. The first Ph.D. with emphasis in environmental engineering was earned in 1957.by Dr. Edward R. Hermann who was Dr. Gloyna's first Ph.D. student. A total of 209 Ph.D. degrees with an environmental engineering specialization were earned during the period 1957 through August 2006. A summary of the number of Ph.D. degrees awarded with emphasis on sanitary engineering, environmental engineering or water resources engineering is presented in Figure 2. On the average approximately 5 Ph.D. degrees were awarded annually through the summer 2006. However 7 Ph.D. degrees were earned in the spring and summer 2006. More civil engineering doctoral candidates earn the Ph.D. degree from The University of Texas at Austin than from any other university in the United States. Historically, most of the Ph.D. graduates entered into academia; however, recent graduates have been employed by environmental engineering consulting firms, governmental research laboratories as post-doctoral fellows, in addition to academia.

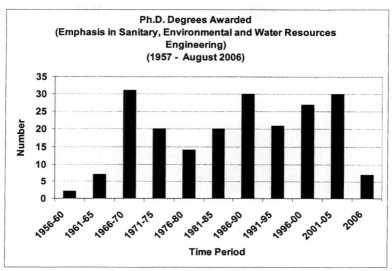

Figure 2. Ph.D. Degrees Awarded at The University of Texas at Austin
(Sanitary, Environmental & Water Resources Engineering Emphasis)
(1956-August 2006)

National Ranking

The Environmental and Water Resources Engineering program at The University of
Texas has developed into one of the premier centers of graduate engineering
education and research in the United States.   The strong and diverse academic
program, excellent accomplishments of the M.S.E. and Ph.D. graduates and a
qualified and respected faculty have been recognized. The U.S. News and World
Report has consistently ranked the Environmental and Water Resources Engineering
graduate program at The University of Texas at Austin among the top 7
environmental engineering graduate programs in the United States. In the 2006 U.S.
News and World Report rankings the UT EWRE program is tied for 2nd place among
environmental engineering graduate programs in the United States.

References
McCaslin, R.B and E.F. Gloyna (1986). *"Commitment to Excellence-One Hundred
Years of Engineering Education at The University of Texas at Austin,"* The
Engineering Foundation of the College of Engineering, The University of Texas at
Austin, Austin, TX

Woolrich, W.R. (1964). *"Men of Ingenuity from beneath the orange tower,"* The
Engineering Foundation of the College of Engineering, The University of Texas at
Austin, Austin, TX

Table 1 Chronological Listing of Faculty
Sanitary, Environmental & Water Resources Engineering
The University of Texas at Austin (1888-2006)

Name	Tenure at Univ of Texas	Current Title at Univ of Texas	Engineering Specialty
T.U. Taylor[3]	1888-1937	Emeritus	General Water Resources
	C.E., Cornell University		
Edward .C.H. Bantel[3]	1901-1952	Emeritus	Hydrology/Sanitary
	C.E., Rensselaer Polytechnic Institute		
Quintin B. Graves[3]	1937-1946	Emeritus	Hydraulics/Sanitary
Carl W. Morgan[3]	1946-1985	Emeritus	Hydrology/Water Resources
	Ph.D., The University of Texas at Austin		
Kenneth W. Cosens[4]	1947-1953		Sanitary
	Ph.D., Michigan State University		
Walter L. Moore	1947-1984	Emeritus	Water Resources
	Ph.D., University of Iowa		
Benjamin B. Ewing[4]	1947-1955		Sanitary/Water Resources
	M.S.E., The University of Texas at Austin		
Earnest F. Gloyna[2]	1947-2001	Emeritus	Sanitary/Water Resources
	D.Eng., Johns Hopkins University		
Ernest W. Steel[3]	1950-1960	Emeritus	Sanitary
Joe O. Ledbetter[3]	1958-2003	Emeritus	Enviornmenal/Air Pollution
	Ph.D., The University of Texas at Austin		
Joseph F. Malina, Jr.[1]	1961-	Professor	Environmentql/Treatment & Design
	Ph.D., The University of Wisconsin, Madison		
Frank D. Masch[4]	1962-1975		Water Resources
	Ph.D., University of California, Berkeley		
Tom D. Reynolds[3]	1963-1965		Environmentql/Treatment & Design
	Ph.D., The University of Texas at Austin		
W. Wesley Eckenfelder[4]	1965-1969		Enviornmental/Treatment
	M.S.E., Pennsylvania State University		
E. Gus Fruh[3]	1966-1979		Enviornmental/WaterQuality
	Ph.D., The University of Wisconsin, Madison		
Patrick Atkins[4]	1968-1973		Environmental/Air Pollution
	Ph.D., Stanford University		
William S. Butcher[4]	1969-1975		Water Resources
	Ph.D., University of Southern California		
Michael J. Humenick[4]	1970-1981		Enviornmental/Treatment
	Ph.D., University of California, Berkeley		
Richard E. Speece[4]	1970-1974		Enviornmental/Treatment
	Ph.D., Massachusetts Institute of Technology		
Neal E. Armstrong[1]	1971-	Professor	Environmental/Water Quality
	Ph.D., The University of Texas at Austin		
Leo R. Beard[2]	1972-1982	Emeritus	Water Resources
	B.S.C.E., California Institute of Technology		
Gerard A. Rohlich[2,3]	1972-1984	Emeritus	Environmental/Water Resources
	Ph.D., The University of Wisconsin, Madison		
Hall H.B. Cooper[4]	1974-1982		Environmental/Air Pollution
	Ph.D., University of Washington, Seattle		
Davis L. Ford[1,2]	1974 -	Adjunct Prof	Environmental/Treatment & Design
	Ph.D., The University of Texas at Austin		
Larry W. Mays[4]	1976-1988		Water Resources
	Ph.D., University of Illinois, Urbana-Champaign		

Table 1 (continued)
Chronological Listing of Faculty
Sanitary, Environmental and Water Resources Engineering
The University of Texas at Austin (1888-2006)

Name	Tenure at Univ of Texas	Current Title at Univ of Texas	Engineering Specialty
Randall J. Charbeneau[1]	1978-	Professor	Water Resources
Ph.D., Stanford University			
Edward R. Holley	1979-2001	Emeritus	Water Resources
Ph.D., Massachusetts Institute of Technology			
Charles A. Sorber	1980-2006	Emeritus	Environmental
Ph.D., The University of Texas at Austin			
Desmond F. Lawler[1]	1980-	Professor	Environmental/Water Treatment
Ph.D., University of North Carolina Chapel Hill			
Howard M. Liljestrand[1]	1980-	Professor	Environmental/Chemistry
Ph.D., California Institute of Technology			
David R. Maidment[1]	1981-	Professor	Water Resources
Ph.D., University of Illinois, Urbana-Champaign			
Ernest T. Smerdon[2,4]	1982-1986		Water Resources
Ph.D., University of Missouri			
Raymond C. Loehr[2]	1985-2003	Emeritus	Environmental/Remediation
Ph.D., The University of Wisconsin, Madison			
Gerald E. Speitel, Jr[1]	1988-	Professor	Environmental/Treatment
Ph.D., University of North Carolina Chapel Hill			
Daene C. McKinney[1]	1990-	Professor	Water Resources
Ph.D.., Cornell University			
Richard L. Corsi[1]	1994-	Professor	Environmental/Indoor Air Quality
Ph.D., University of California, Davis			
Kerry A. Kinney[1]	1996-	Assoc Prof	Environmental/Air Pollution
Ph.D., University of California, Davis			
Lynn E. Katz[1]	1998-	Assoc Prof	Environmental/Chemistry
Ph.D., University of Michigan			
Spyri don A. Kinnas[1]	2000-	Professor	Water Resources
Ph.D., Massachusetts Institute of Technology			
Ben R. Hodges[1]	2000-	Asst Prof	Water Resources
Ph.D., Stanford University			
Jeffrey A. Siegel[1]	2002-	Asst Prof	Environmental/Indoor Air Quality
Ph.D., University of California, Berkeley			
Danny Reible[1,2]	2004-	Professor	Environmental/Remediation
Ph.D., California Institute of Technology			
Mary Jo Kyrisits[1]	2004-	Asst Prof	Environmental/Microbiology
Ph.D., University of Illinois, Urbana-Champaign			

[1] Current active faculty

[2] Elected to National Academy of Engineering (NAE)

[3] Deceased

[4] No longer affiliated with The University of Texas at Austin on the EWRE faculty

History and Impacts of Levees in the Lower Rio Grande Valley

by John N. Furlong, John Ivey, Joe Barrow, Mike Moya[1], and Ralph O'Quinn[2]

[1] Halff Associates: Dallas & Fort Worth, 8616 Northwest Plaza Drive, Dallas, TX 75225, and
[2] Cobblestone Engineering Inc., Harlingen, TX

I. Introduction

The Lower Rio Grande Valley roots start with early Spanish and Mexican-American settlements along the Rio Grande River in the mid-1700's. It took a lot of grit and determination to live and work in the Valley. Early Valley history is characterized by border wars, politics, and agricultural developments and misappropriated public money. Some of the early drama of the Valley is captured as Texas Online traces the history of one such county – Hidalgo County:

1528 Cabeza de Vaca shipwreck

1638 – Jacinto Garcia de Sepulveda Expedition crossed river to Texas coast in search of Dutch sailors

- 1687 – Second expedition of Alonso De Leon in search of Ft. St. Louis followed river route
- 1747 – Miguel de la Garza Falcon looked for suitable land to start settlement. No luck.
- 1749 – 1752 – Jose de Escandon set out to establish four towns: Reynosa-1749 (which was originally located across the river from the site of present-day Peñitas), Camargo-1749, Mier-1750, and Revilla (now Guerrero) - 1752.
- 1750–1800 - Settlers from these colonies later crossed the Rio Grande and settled the northern banks of the river.
- Early settlements were primarily ranches – land grants and small communities started with permission from the Spanish authorities in Mexico.

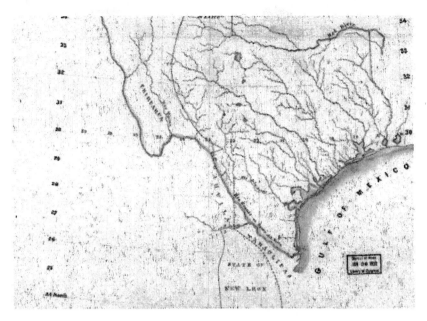

Figure 1 - Sketch of Texas with the boundaries of Mexican States as shown on General Austin's map of Texas published by R. S. Tanner, 1839.

- Early 1800's - By 1836 area farmers had a thriving economic base that allowed them to export their cattle and cattle by-products into Mexico. **Goods were moved by wagon and mule trains, whose owners were so organized that they kept boats off the Rio Grande until after 1840.** With the outbreak of the Texas Revolution[qv] the area became disputed territory, Mexico considered it part of Tamaulipas, and Texas claimed it as part of its southern border. During the Mexican War,[qv] Zachary Taylor[qv] laid out the Old Military Road[qv] to supply his men in northern Mexico. In 1849 the area became a popular stopping point for goldseekers from the United States on their way to California. By 1850 about thirty-nine ranches were in operation in what later became Hidalgo County. Mexico was the main market for goods from the area. Residents grew a variety of fruits and vegetables, including squash, citrus fruit, and corn.

- 1850-75 - Hidalgo County was formed in 1852 and named for Miguel Hidalgo y Costilla,[qv] who gave the "cry for Mexican independence" from Spanish rule. By 1852 the county had between forty and forty-five ranches. **As land was** parceled out from one generation to the next the ranches located along the river developed into villages. In this way, ranches gave rise to the communities of La Habitación, Relampago, and Peñitas. In 1852 La Habitación was renamed Edinburgh and made county seat. The first county court convened on September 2, 1852, and as its first act issued licences to ferries at Hidalgo, San Luis, Peñitas, and Las Cuevas. County residents were isolated from each other, however, and from the population center of Brownsville in neighboring Cameron County. Because of their sense of neglect by state and federal governments, residents adopted the name "Republic of Hidalgo." Isolation and ineffective law enforcement led to general chaos and lawlessness, mostly in the form of cattle raids and shootouts. As early as December 28, 1862, armed Mexican bandits crossed into Los Ebanos, captured a Confederate wagon train, and killed three teamsters.

- 1876-1890 – Hidalgo County had become a haven for outlaws from both sides of the river by the middle of the nineteenth century. Politically it had become a battleground, as various groups vied for dominance of county politics. By 1880 the population was 4,347, and all except women and the 114 African Americans[qv] were fair game for the parties looking for votes. Not until 1882, when John Closner[qv] was elected deputy sheriff, was control over cattle rustlers achieved. Closner became sheriff in 1890 and shortly afterward, under the protection of James B. Wells,[qv] became the county's political boss. During his rule he brought peace to the county and was seen as such an effective leader that he was nicknamed the "father" of Hidalgo County. In the process, however, he made many enemies. During the 1890s his rivals tried to have him assassinated twice and brought a ranger investigation against him. He was accused of mistreating prisoners, and he later admitted that he could have gone a little too far in pressuring suspects to confess to crimes.

- 1890–1910 – Despite political turmoil and cattle rustling, the county population grew to 6,534 by 1890. In 1886 Edinburgh was washed away by a severe flood, after which it was moved to another flood-prone site about two miles north of the river. The county population was estimated at 6,837 in 1900. The first attempt at growing cane on a large scale was made in 1883 by John Closner, who established a plantation and mill near the site of present-day Pharr. Attempts to irrigate rice were unsuccessful, but citrus fruits and vegetables were produced on a commercial basis starting around 1907, when W. A. Fitch planted a commercial-scale grapefruit orchard near Mercedes. Chapin, a community established in 1908, was soon made county seat and renamed Edinburg. The old county seat, Edinburgh, was moved away from the river and renamed Hidalgo. With the introduction of the railroad and the influx of settlers wishing to establish farms during the first decade of the twentieth century, the county's economic base shifted toward farming. The primary crops were corn and cotton. The population was estimated at 13,728 in 1910. In 1911 the San Benito and Rio Grande Valley

Railway made junction with the St. Louis, Brownsville and Mexico Railway at San Benito. The Texas and New Orleans built into the Valley in 1927.

- 1910-1930 – During the first decade of the twentieth century Closner and his Democratic machine ran unopposed, **Closner's reign ended in 1918, when an audit revealed that as county treasurer he had misappropriated $150,000 from the county, drainage districts, and the school district.**

Hydrologic Setting

The Rio Grande drainage basin begins near the U.S. Continental Divide in southern Colorado and extends through New Mexico, Texas and Mexico to the Gulf of Mexico. The portion of the River between El Paso, Texas, and the Gulf of Mexico, a distance of 1,250 river miles, forms the international boundary between the United States and Mexico. The drainage basin comprises an area of 355,500 mi^2, with the water-contributing area totaling 176,000 mi^2. The remaining portion of the basin drains into internal closed sub-basins, and does not contribute water to the River. About half of the River's water-contributing areas (about 89,000 mi^2) is located in the United States, with the Texas portion encompassing about 54,000 mi^2 of the total.

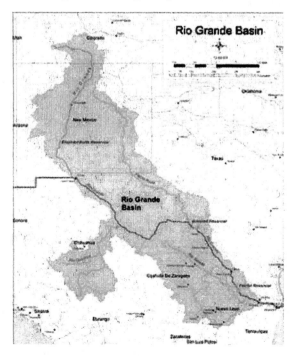

Figure 2-The overall drainage basin, political subdivisions, and major tributaries of the Rio Grande River.

The Pecos and Devils Rivers are the principal Texas tributaries of the Rio Grande, both flowing into Amistad Reservoir, about 600 river miles from the River's mouth. The largest Mexican tributaries draining to the river are the Rio Conchos, Rio Salado and Rio San Juan. The Rio Conchos drains about 26,000 mi^2, flowing into the Rio Grande near Presidio, Texas, about 350 river miles above Amistad Reservoir. The Rio Salado drains an area of about 23,000 mi^2, discharging directly into Falcon Reservoir, about 275 river miles upstream from the River's mouth. The Rio San Juan drains about 13,000 mi^2, entering the River about 36 river miles below Falcon Dam near Rio Grande City, Texas.

Most of the Rio Grande drainage basin is comprised of rural, undeveloped land used principally for farming and ranching. The major Texas--Mexico paired urban areas on the River are El Paso--Ciudad Juarez, Del Rio--Ciudad Acuna, Eagle Pass--Piedras Negras, Laredo--Nuevo Laredo, McAllen-Reynosa, and Brownsville--Matamoros. Substantial quantities of surface water are diverted from the River to meet municipal and agricultural demands in Texas and Mexico, with much of is demand being in the Lower Rio Grande Valley. Although most of the cities in the Valley are located outside its contributing watershed, the River serves as their primary water supply.

The Rio Grande Basin is divided into three more-or-less distinct regions. The upper Rio Grande originates in the Colorado/New Mexico mountain ranges (Southern part of the Rocky Mountain chain) and extends south to El Paso, Texas. The middle region starts south of El Paso and extends to the Falcon Dam south of Laredo, Texas. The lower region starts below Falcon Dam and extends to the Gulf of Mexico. The focus of this paper is on flood control and levees in the Lower Rio Grande Valley.

The Lower Valley is characterized by extensive irrigated agriculture of significant economic importance for the region. Most of the water diverted from the River in the lower Rio Grande Valley is not returned to the River as irrigation tailwater or treated wastewater effluent because of the land's natural slope away from it. The return flows are instead discharged into interior manmade drainage channels and floodways that ultimately drain into the Laguna Madre estuary, and thence into the Gulf of Mexico. Flooding along the Lower Rio Grande Valley is controlled to a large extent by two major dams on the mainstem of the River: Amistad and Falcon, forming the two major international reservoirs on the River's mainstem. Amistad and Falcon Reservoirs provide controlled water storage for over eight million acre-feet of water owned by the two countries. Of this quantity, 2.25 million acre-feet (one acre-foot is equivalent to about 326,000 gallons) are allocated for flood control and 6.05 million acre-feet are reserved for silt and conservation storage (water supply). Falcon Reservoir, completed in 1953, is considered to be the "lowest major international dam or reservoir" on the River, in accordance with the provisions of the 1944 Treaty. The United States owns 1.56 million acre-feet (58.6%) of the reservoir's silt and conservation and storage, with Mexico owning the remaining 1.10 million acre-feet.

Amistad Reservoir, completed in 1968, is located on the River above Del Rio, Texas. The United States controls 1.77 million acre-feet (56.2%) of the total conservation storage capacity, with Mexico owning the remaining 1.38 million acre-feet. Anzalduas Dam, completed in 1960 and located near Mission, Texas, is used as a storage and flow regulation facility for partially controlling the United State's share of the water in this reach of the lower portion of the River. It also enables diversions into Anzalduas Canal, Mexico's main water delivery canal. The reservoir has a total storage capacity of 15,000 acre-feet at its normal maximum operating level, with between 3,037 and 4,214 acre-feet available as conservation storage for the United States.

Mexico has constructed reservoirs on its Rio Grande tributaries that have a combined water storage capacity larger than the total available to it in Amistad and Falcon Reservoirs. This water is used for municipal, industrial and irrigation purposes in the vicinity of the reservoirs and downstream along the tributaries. Pertinent features of the reservoirs located on the Mexican tributaries are summarized in Table 1. Much of the Mexican reservoir development has occurred in the Rio Conchos Basin in the State of Chihuahua, which flows into the Rio Grande upstream of Amistad Reservoir. The 1944 Treaty identifies the Rio Conchos as one of the six Mexican tributaries from which the United States is allocated a portion of the inflows to the Rio Grande. The combined conservation storage capacity of all the major Mexican tributary reservoirs is about 6,240,000 acre-feet (Table 1), equivalent to about 2.5 times Mexico's total available conservation storage capacity in Amistad and Falcon reservoirs. Mexico also is constructing a new reservoir (Las Blancas) on the lower reach of the Rio San Alamo, one of the Mexican tributaries flowing into the River below Falcon Dam. Its use apparently will be to capture floods and convey them by canal to the existing Marte R. Gomez Reservoir on the Rio San Juan.

Table 1- The primary surface water reservoirs on the Rio Grande River.

River basin/name	River	State	Year Completed	Storage Capacity Million m³	Acre-feet
Rio Conchos Basin:					
La Boquilla	Rio Conchos	Chihuahua	1916	2,903	2,353,501
La Colina	Rio Conchos	Chihuahua	1927	24	19,538
Francisco I. Madero	Rio San Pedro	Chihuahua	1949	348	282,128
Chihuahua	Rio Chuviscar	Chihuahua	1960	26	21,079
Luis L. Leon	Rio Conchos	Chihuahua	1968	356	288,614
San Gabriel	Rio Florida	Durango	1981	255	206,732
Pico de Aquila	Rio Florida	Chihuahua	1993	50	40,536
Rio Conchos Basin total reservoir storage capacity				*3,962*	*3,212,127*
Rio San Diego Basin:					
San Miguel	Rio San Diego	Coahuila	1935	20	16,214
Centenario	Rio San Diego	Coahuila	1936	26	21,322
Rio San Diego Basin total reservoir storage capacity				*46*	*37,536*
Rio San Rodrigo Basin:					
La Fragua	Rio San Rodrigo	Coahuila	1990	45	36,482
Rio San Rodrigo Basin total reservoir storage capacity				*45*	*36,482*
Rio Salado Basin:					
Venustiano Carranza	Rio Salado	Coahuila	1930	1,385	1,112,838
Laguna de Salinillas	Rio Salado	Nuevo Leon	1931	--	--
Rio Salado Basin total reservoir storage capacity				*1,385*	*1,112,838*
Rio San Juan Basin (1):					
Rodrigo Gomez	Rio San Juan	Nuevo Leon	1963	41	33,239
El Cuchillo	Rio San Juan	Nuevo Leon	1993	1,123	910,512
Marte R. Gomez	Rio San Juan	Tamaulipas	1943	1.097	889,271
Rio San Juan Basin total reservoir storage capacity				*2,261*	*1,833,023*
Total Mexican Reservoir Storage Capacity				*7,699*	*6,242,006*

(1) Flow from these reservoirs is dedicated to Mexico by the 1944 Treaty:

Mexico's share of water conservation storage in Rio Grande/Rio Bravo international reservoirs					
River basin/name	River	State	Year completed	Storage capacity million m³	Acre-feet
Rio Grande Basin:					
Falcon Reservoir	Rio Grande	Tamaulipas	1953	1,355	1,098,674
Amistad Reservoir	Rio Grande	Coahuila	1969	1,703	1,380,278
Total American Reservoir Storage Capacity				*3,058*	*2,478,952*

An understanding of the hydrologic setting of the Rio Grande River is important in understanding the water available downstream for diversion for irrigation and water supply.

II. Early Beginnings – Archeological Record

The Brownsville-Barril complex or culture is the name that archeologists have given to the remains left by the little-known Indian groups who occupied the Rio Grande Delta at the extreme southern tip of Texas during the time period of A.D. 1100-1700. The low-lying tropical delta area was one of the last parts of northeastern Mexico to be explored and settled by the Spanish. They did not turn their attention to the Rio Grande Delta until 1747, about 150 years after they first established towns in Nuevo Leon and Coahuila. The mid-eighteenth century accounts suggest that as many as 50 named Indian groups lived in the Rio Grande Delta.

In exchange, the Brownsville-Barril peoples received exotic items including pottery, jade, and obsidian that are rarely found in Texas beyond the Rio Grand Delta. In fact this is one of the most interesting archeological questions about the area: What was the nature of trade and contact between these nomadic delta peoples and the much more sophisticated cultures to the south?

The native Americans of the Rio Grande Delta were also distinctive in the manner in which they buried their dead. Individuals were buried in tightly flexed positions, and graves were located away from living areas. The deceased was accompanied by offerings such as shell beads and pendants, animal and human bone awls, and bone beads or tubes used as jewelry. With some burials, red pigment powder was strewn over the burial. Others included Mesoamerican goods such as prehistoric pottery vessels from the Tampico, Mexico region, obsidian (volcanic glass) arrow points from sources in central Mexico, and greenstone (jadeite) jewelry, also from Mexico.

Much of what is known about the Brownsville-Barril complex is due to the efforts of a civil engineer and draftsman from Brownsville, Texas. From 1917 to 1941, Andrew Eliot Anderson Much of what is, took extensive notes, drafted maps, and collected distinctive artifacts from archeological sites of the Rio Grande Delta and Laguna Madre of coastal south Texas and northeastern Mexico. After World War II, the area was drastically transformed by land clearing, agricultural modification, and urban development. Today most of the sites that Anderson visited have been destroyed. The A. E. Anderson collection, now housed at the Texas Archeological Research Laboratory, is a crucial research resource.

The Rio Grande Delta of today bears scarce resemblance to its appearance prior to the twentieth century. The great river has been reduced to a trickle by upstream dams and heavy agricultural demand. The fertile soils and mix of marsh, waterways, and raised areas have been homogenized—the smaller waterways filled in, the clay dunes flattened, and the area covered by expanses of agricultural fields, orchards, and urban areas. The majority of the archeological record has been destroyed, particularly the most fragile and most visible materials that were once common on and near the surface. The best potential for learning more about the Brownsville-Barril complex is a combination of renewed ethnohistoric research—combing the Spanish archives for eyewitness accounts of the native peoples of the Rio Grande Delta—and analyses of the existing collections and notes.

Figure 3 - Soldiers working on levees after Flood of 1916 on Rio Grande River near Brownsville.

Figure 4 - Soldiers working on levees after Flood of 1916 on Rio Grande River near Brownsville.

A. Flood Events on the Rio Grande River

Major flood events on the Lower Rio Grande River occurred in 1865, 1886, 1904, 1909, 1914, 1916, 1922, 1948, 1954, 1967, and 1988 which caused considerable damages in a region generally east of Mission in Hidalgo, Cameron, and Willacy counties. In 1925 and 1926, Hidalgo County, in cooperation with Cameron County, began the first flood protection project in the lower valley, consisting of a river levee and interior floodways.   The interior floodways consisted of the Mission and Hackney inlets and the Main Floodway and were formed by locating levees along natural overflow channels. Major recent hurricanes hitting the Lower Valley include Hurricane Beulah in 1967 and Hurricane Gilbert in 1988. Both of these events caused severe flooding and damages to residential and commercial properties in the Lower Rio Grande Valley. Table 2 below shows some of the significant peak flows at USGS and IBWC gaging stations along the Rio Grande River.

USGS station no.	USGS station name	Latitude	Longitude	Contributing drainage area (mi²)	Maximum known peak discharge Date	Maximum known peak discharge Discharge (ft³/s)
08377500	Rio Grande at Langtry, Texas	29°48'00"	101°34'00"	81,429	00/00/1922	204,000
08377600	Rio Grande tributary near Langtry, Texas	29°48'17"	101°29'01"	32	04/17/1969	141
08407800	Delaware River tributary near Orla, Texas	31°55'46"	104°28'52"	38	08/21/1966	1,700
08411500	Salt Screwbean Draw near Orla, Texas	31°52'40"	103°56'50"	464	10/02/1955	40,600
08424500	Madera Canyon near Toyahvale, Texas	30°52'04"	103°58'09"	53.80	09/26/1932	5,120
08431700	Limpia Creek above Fort Davis, Texas	30°36'48"	104°00'04"	52.40	06/19/1984	8,610
08431800	Limpia Creek below Fort Davis, Texas	30°40'52"	103°47'30"	227	00/00/1932	14,200
08433000	Barrilla Draw near Saragosa, Texas	30°57'28"	103°27'33"	612	09/08/1981	2,200
08434000	Toyah Creek below Toyah Lake near Pecos, Texas	31°21'00"	103°24'00"	3,709	08/07/1940	5,850
08435620	Alpine Creek at Alpine, Texas	30°21'06"	103°40'00"	18.10	09/20/1974	3,210
08435660	Moss Creek near Alpine. Texas	30°20'10"	103°38'24"	11.30	09/20/1974	3,760
08435700	Sunny Glen Canyon near Alpine. Texas	30°22'52"	103°44'08"	29.70	09/21/1972	570
08435800	Coyanosa Draw near Fort Stockton, Texas	31°02'27"	103°08'15"	1,182	06/15/1967	12,600
08436800	Courtney Creek tributary near Fort Stockton, Texas	31°00'28"	103°04'20"	1.14	05/29/1972	116
08444400	Three Mile Mesa Creek near Fort Stockton, Texas	30°50'16"	102°50'26"	1.04	07/19/1973	350
08447020	Independence Creek near Sheffield, Texas	30°27'07"	101°43'58"	763	09/20/1974	78,100
08447400	Pecos River near Shumla, Texas	29°50'00"	101°23'00"	35,162	06/28/1954	948,000
08449000	Devils River near Juno, Texas	29°57'48"	101°08'42"	2,730	00/00/1954	393,000
08449400	Devils River at Pafford Crossing near Comstock, Texas	29°40'35"	101°00'00"	3,961	09/18/1974	250,000
08449470	Rough Canyon tributary near Del Rio, Texas	29°35'50"	100°51'51"	7.90	09/11/1972	8,540
08449500	Devils River near Del Rio, Texas	29°29'00"	101°00'00"	4,185	09/01/1932	597,000
08449600	Evans Creek tributary near Del Rio, Texas	29°33'00"	101°04'58"	.39	08/11/1971	313
08450500	Devils River near Mouth, Texas	29°28'10"	101°03'25"	4,305	09/21/1964	122,000
08450900	Rio Grande below Amistad Dam near Del Rio, Texas	29°25'00"	101°02'00"	123,143	06/00/1954	1,158,000
08452500	Rio Grande near Del Rio, Texas	29°20'00"	100°56'00"	123,303	06/28/1954	1,140,000
08453000	San Felipe Creek near Del Rio, Texas	29°19'55"	100°53'20"	46	06/14/1935	45,000
08453100	Zorro Creek near Del Rio, Texas	29°19'52"	100°49'54"	10	06/13/1935	2,000
08454500	East Perdido Creek near Brackettville, Texas	29°20'50"	100°34'32"	3.39	08/12/1971	630
08455000	Pinto Creek near Del Rio, Texas	29°08'45"	100°43'05"	249	06/24/1948	186,000
08458000	Rio Grande at Eagle Pass, Texas	28°42'50"	100°30'25"	127,312	06/00/1865	1,236,000
08458700	Rio Grande at San Antonio Crossing, Texas	28°21'00"	100°18'00"	129,226	06/29/1954	912,000
08459000	Rio Grande at Laredo, Texas	27°29'45"	99°29'25"	132,578	00/00/1865	950,000
08459600	Arroyo San Bartolo at Zapata, Texas	26°55'39"	99°17'20"	.61	05/11/1969	620
08460500	Rio Grande near Zapata, Texas	26°52'00"	99°18'00"	163,344	09/04/1932	261,000
08462500	Rio Grande at Roma, Texas	26°24'00"	99°01'00"	166,464	00/00/1865	630,000
08464700	Rio Grande at Fort Ringgold, Texas	26°22'05"	98°48'20"	174,362	00/00/1865	590,000
08466100	Rio Grande tributary near Rio Grande City, Texas	26°18'58"	98°39'45"	1.20	09/22/1967	125
08466200	Rio Grande tributary near Sullivan City, Texas	26°17'12"	98°35'16"	.40	04/27/1972	195

Table 2 - Significant Peak Flows in the Rio Grande Basin – Note the 1865 flood event appears to be the flood of record at several downstream gaging stations.

B. Role of International Boundary and Water Commission (IBWC)

The agency originally known as the International Boundary Commission (IBC) was created by the Convention of 1889. It eventually became known as the IBWC with the signing of the 1944 Treaty, which provided for both a United States Section and a Mexican Section. The IBWC is the agency tasked with applying the boundary and water treaties between the two countries in a manner which "... *benefits the social and economic welfare of the peoples on the two sides of the boundary and improves relations between the two countries.*" Specific IBWC tasks include: accounting for and distributing international waters of the Rio Grande; and overseeing the construction, maintenance, and operation of all infrastructure, including reservoirs, dams, hydroelectric energy-generation facilities, floodways, and levees downstream of Caballo Reservoir in New Mexico. The international boundary between the United States (U.S.) and Mexico is over 1,952 miles in length, with the Rio Grande encompassing 1,254 miles of that total. Today, the boundary is characterized by fifteen pairs of sister cities sustained by agriculture, import-export trade, service and tourism, and in recent years, by a growing manufacturing sector. The entire borderlands' population (i.e., the entire 1,952 mile corridor encompassing cities' populations on both sides of the border) was estimated to be 10.6 million in 1995 (IBWC 2004a).

C. Historical Flooding of Lower Rio Grande Valley and Development in the Estuary Reach

An understanding of drainage in the Lower Rio Grande Valley is necessary to understand the issues involved with competing interests of agriculture, transportation, housing, and commercial development. The discussion below explains some history and flooding in the Lower Rio Grande Valley.

The Lower Rio Grande Valley was hit by Hurricane Beulah in September 1967 causing more damage than any other recorded storm. In McAllen a rainfall total of 16.08 inches was recorded. The return interval for this event was about 125 years. Hurricane Beulah caused widespread flooding in McAllen, much of which was caused by the overtopping of a flood channel levee. The Mission Inlet to the International Boundary and Water Commission's (IBWC) Main Floodway, which runs just south of McAllen, had a levee that was overtopped causing severe flooding in the vicinity of the McAllen airport. Since 1967 the Mission Inlet has been abandoned as part of the floodway system, but it is still used to carry local stormwater runoff. The size of the flood channel is more than sufficient to carry any local flows it receives without danger of overtopping.

The principal flood problem in the McAllen area is the flatness of the terrain. Because of the low relief, man-made barriers such as roadway and railroad embankments, or elevated irrigation canals, severely limit the removal of stormwater. In addition, small natural depressions exist that pond water to their rim elevations. The only area of widespread flooding occurs just west of FM 1926 between the Mission Inlet and the Banker Floodway. Here the terrain is very flat with numerous low areas. This condition is further aggravated by its location between two levees to the north and south and an elevated irrigation canal to the east. The absence of topographic relief and natural drainage features are the major factors that contribute to flooding. Land slopes on the order of one foot per 1500 feet and drain ditch bottom slopes of one foot per mile are common. Flat terrain coupled with soils having low infiltration capacities combine to make shallow ponding of rainfall runoff the most common type of flooding in Hidalgo County. While overland runoff moves very slowly toward manmade drainage ditches, numerous obstructions such as elevated canal levees, dikes, roadways, and railroad embankments tend to impede the flow.

Ponding can be very widespread; during Hurricane Beulah, thousands of acres of land were inundated in Hidalgo county. This ponding was usually shallow--less than one foot--but in some areas it persisted for months after the storm had passed.

There is an extensive system of manmade drainage ditches throughout the County, but the system has its limitations. Because of the flat terrain, it takes several days for stormwater to move overland to the drains. Once in the drains, water continues to move slowly, since the bottom slopes of the drains are fairly small, thus limiting their capacity. Also, undersized culverts though roadway embankments restrict the flow of the drains and create excessive backwater.

Another problem relates to the termination of drains. Most drainage ditches throughout the county discharge into the IBWC floodway system through gated structures in the floodway levees. This operating procedure is functional only when the floodway system is not being used. Otherwise, during flood periods on the Rio Grande when excess floodwaters are diverted through the IBWC floodway system, water levels inside the floodway levees inundate the drain ditch discharge structures, and the drain outlet gates must be closed to prevent surcharging. With the outlet gates closed, ponding occurs behind the floodway levees; this causes additional flooding problems.

The Rio Grande forms the southern boundary of Hidalgo County. Although there is a short unleveed section of the river on the extreme western end of the county, most of the river is confined between flood control embankments. The leveed portion extends from the Town of Penitas east to the Cameron County line. West of Penitas to the Starr County line no levees exist. This unleveed area is sparsely populated and is used mainly for agricultural purposes. Also, a steep erosional escarpment exists on the northern edge of the Rio Grande's flood plain. This bank tends to contain the river's flood flows.

Table 1. Selected Summary Characteristics of IBWC Project Areas, U.S. Side, 2004.

	Project Area (Economic Reach)				
	Rio Grande Canalization	Rio Grand-Rectification	Presidio Valley Flood Control	Lower Rio Grande Flood Control	Total
	U.S. Side Only				
Flood Information					
flood design	100-Year	100-Year	25-Year	500-Year	--
freeboard	2 feet	2 feet	4 feet	3 feet	--
designed-flood flow (CFS)	22,200 /14,000	11,000	3,600 / 42,000	250,000	--
- at location	at Leasburg Dam / at American Dam	El Paso, TX.	above / below Rio Conchos	Rio Grande City, TX.	--
River Floodway					
- total miles of levee	130	93	15.18	102	340.18
- miles at risk of being overtopped	10	12	1.25	38	61.25
- miles subject to encroachment	60	38	1.25	64 [a]	163.25
Interior Floodway [b]					
- total miles of levee	0	0	0	172	172
- miles at risk of being overtopped	0	0	0	2	2
- miles subject to encroachment	0	0	0	24 [a]	24
Area In Revised FEMA Flood Plain (acres)					
- Agriculture	2,484	2,356	764	75,645	81,249
- Residential	1,836	2,643	320	3,237	8,036
- Commercial	65	2,759	0	605	3,429
- Industrial	0	32	0	0	32
Total [last four rows only]	4,385	7,790	1,084	79,487	92,746

Sources: IBWC 2004c, Jim Robinson and Albert Moehlig with the IBWC, and data calculations of project personnel.
[a] Note for the LRGFC project, values based on river levees having 3-foot of freeboard and floodway levees having a 2-foot freeboard.
[b] These values are summed for the left and right levees (i.e., with a downstream flow).

Table 3 – Characteristics of the Lower Rio Grande Flood Control Project

1. Lower Rio Grande Flood Control Project

The Lower Rio Grande Flood Control (LRGFC) project extends 158 miles along the main channel of the Rio Grande from Penitas in Hidalgo County, Texas to a point 28 miles from the Gulf of Mexico. The USIBWC infrastructure inventory in the Lower Rio Grande Flood-Control (LRGFC) project includes some 274 miles of river and interior (i.e., off-river) floodway levees. Of this total, a combined 40 miles are estimated to be overtopped (on the U.S. side of the river and along the interior floodway) by water flow associated with a 500-year flood event (Moehlig), while 88 miles are subject to encroachment (IBWC 2003). The LRGFC project's purpose is to provide flood protection to urban, suburban, and high-value agriculture production in the area. The LRGFC project is designed to protect the area from waters associated with a 500-year flood (IBWC 2004c). For this project area, the IBWC defines such flooding as a flow rate of 250,000 CFS at Rio Grande City, TX.. Based on a 1932 IBWC report on flood control in the Lower Rio Grande, the U.S. and Mexico agreed on a coordinated plan to prevent the individual countries' flood-protection actions from exacerbating floods on the other side of the border. The U.S. portion of this project, covering Cameron, Hidalgo, and part of Willacy Counties, was authorized under Title II of the National Industrial Recovery Act (Act) of June 13, 1933 and the Act of August 19, 1935. The project involved the construction of a system of river levees and leveed-interior floodways in each country that would be maintained by the respective IBWC Section. Two diversion dams, Anzalduas and Retamal, divert floodwaters into the U.S. and Mexican interior floodways, respectively. The project's original flood-capacity design of 187,000 CFS (measured at Penitas, Texas) was increased to 250,000 CFS (measured at Rio Grande City, Texas) following extensive damages caused by Hurricane Beulah in 1967.

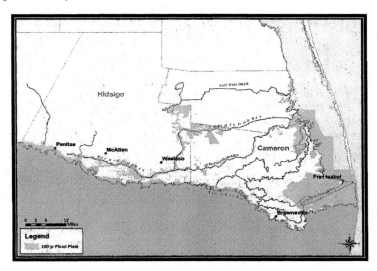

Figure 5 - Location of the Lower Rio Grande Flood Control Project, 2004.

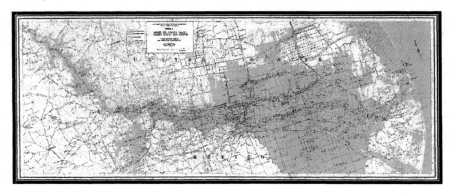

Figure 6 – Lower Rio Grande Flood Control Project showing irrigated areas, urban areas, and floodway and major outfalls

Although the Rio Grande/Rio Bravo flows directly into the Gulf of Mexico, its final 30 miles (essentially that portion below Brownsville) is a tidal river system. In combination with the offshore portion of the Gulf of Mexico directly influenced by the River's freshwater discharge plume, these two components comprise the River's estuary system. Because of the generally low rainfall and small river flows, the Rio Grande river system is stratified, with freshwater at the surface down to river mile 10, and a saltwater wedge on the bottom extending upriver about 17 miles (average water conditions during the 1992 – 1997 period). Tidal exchange generally dominates the flow and mixing regime in this area, except when locally-heavy rainfalls occur in the late summer-early fall monsoon period. The flows in the lower portion of the River into the Gulf of Mexico are controlled by flows past the Anzalduas Dam at Mission, Texas. These downstream flows, in turn, depend on water diversions made for irrigation, industrial, and municipal uses throughout the lower basin of the River.

The precarious nature of the Rio Grande estuary became especially evident in February, 2001, when the mouth of the River at Boca Chica was blocked by a sandbar, because of low-flow conditions resulting from the severe drought the lower River basin had been experiencing since 1995. The rivermouth remained closed until September, 2001, when it was dredged open temporarily by the International Boundary and Water Commission. Although the excavated rivermouth channel remained open months as a result of tidal water exchanges, it was again closed with silt. The mouth of the River remained closed until September 2002, when higher tides and slightly-increased inflows from rainfall events partially opened it. An analysis of the biological impacts to this estuary from the closing of the rivermouth indicated that the most important function of the freshwater inflows from the lower River is to provide reduced salinity habitat for post-larval and juvenile marine species to complete their life cycles. Based on preliminary TPWD data, lacking a means of ingress and egress to this habitat, the production of some aquatic species has decreased (Randy Blankinship, personal communication).

The return flows of water diverted from the Rio Grande/Rio Bravo currently pass down the Arroyo Colorado, emptying into the Laguna Madre, the estuary north of the River's estuary. If some of this diverted water were re-routed from the Arroyo Colorado via pipeline to the tidal portion of the River, it might be sufficient to satisfy the minimal target freshwater inflow requirements of the estuary.

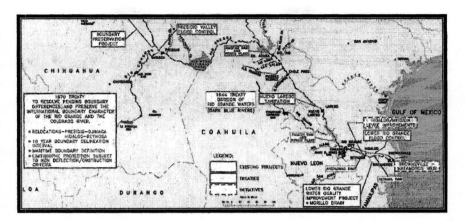

Figure 7- Major Projects, Treaties and Initiatives in the Lower Valley.

2. Model Studies of Lower Rio Grande Valley by Others

Over the last 20 years several studies investigating the flooding characteristics of the Lower Rio Grande Valley have been conducted. The TWDB has partially funded with local participation several Flood Protection studies for the cities of:

*   1995 - Donna – Rust-Lichliter-Jamieson – Martha Juch
*   1996 - McAllen – Mission – Halff Associates – Mike Moya
*   1997 - Hildalgo County – Turner, Collie & Braden – Allen Potok
*   1999 - Roma – Perez/Freese & Nichols, Inc. – Tony Reid
*   2004 - Raymondville – MGM Engineering Group – Flood Mitigation Plan
*   2006 - Brownsville – Ambiotic, Rice Univ. – Carlos Marin

Other drainage studies completed for Drainage Districts, IBWC or Bureau of Reclamation have also modeled various streams and reaches in the Lower Rio Grande Valley. Several of these studies are listed as references at the end of this paper.

The hydrology used in many of these studies included the SCS or NRCS methods for flow development using techniques in HECHMS. Hydraulic routing techniques vary depending on channel conditions and ponding effects. Water surface elevations have been determined with either tidal models along the coast, or HECRAS on more riverine channels.

One of the most significant model studies was a hydraulic model completed by the IBWC as a result of litigation in 2003 entitled, "Hydraulic Model of the Rio Grande and Floodways within the Lower Rio Grande Flood Control Project:"

"The purpose of the present study is to determine the capability of the LRGFCP to convey the design flood flows, under existing vegetation conditions, at water surface

60   ENVIRONMENTAL AND WATER RESOURCES

elevations that will not encroach on the proposed 3 ft. freeboard for the Rio Grande levees and 2 ft. freeboard for the floodway levees. The occurrence of such encroachment at any point would indicate a diminished capability of the LRGFCP to provide adequate flood protection at that point."

The study concluded that: "From a point of view of flood risk generated from encroachment on the recommended 3ft. (0.9 meters) levee freeboard in the Rio Grande, it is concluded that in general the LRGFCP does not provide adequate flood risk protection along its full length for the 100-year flood of 250,000 cfs (7,080 cms) at Rio Grande City."

III. Competing Needs – Politics and Water

A. American Waters vs. Mexican Waters

Several interests are competing for irrigation water and domestic water usage. The regional water planning group M includes 8 counties along the Rio Grande River (Figure 8).

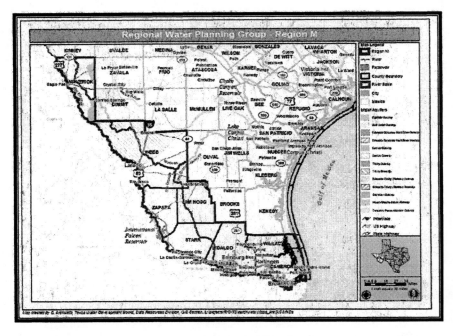

Figure 8 – Regional Planning Group M – Including the Lower Rio Grande Valley area.

Water usage projections for Planning Group M is shown in Table 4. As can be seen the demands for municipal water will nearly triple in the next 50 years. The population in the Lower Valley region is expected to triple as well.

2006 Regional Water Plan
Population and Water Demand Projections Summary for Region M

Regional Total Projection

	D2000	D2010	D2020	D2030	D2040	D2050	D2060
Population	1,236,246	1,581,207	1,973,188	2,401,223	2,854,613	337,618	3,826,001
Irrigation (AF/YR)	1,209,647	1,163,633	1,082,231	981,749	981,749	981,749	981,749
Livestock (AF/YR)	5,817	5,817	5,817	5,817	5,817	5,817	5,817
Manufacturing (AF/YR)	6,208	7,509	8,274	8,966	9,654	10,256	11,059
Mining (AF/YR)	3,869	4,186	4,341	4,433	4,523	4,612	4,692
Municipal (AF/YR)	226,536	279,633	338,716	403,511	472,632	547,747	625,743
Steam Eelctric (AF/YR)	6,780	13,463	16,864	19,716	23,192	27,430	32,598
Total Water Demand (AF/YR)	1,458,857	1,474,241	1,456,243	1,424,192	1,497,567	1,577,611	1,661,658

Region M Population Projection by County

	D2000	D2010	D2020	D2030	D2040	D2050	D2060
Cameron	335,227	415,136	499,618	586,944	673,996	761,073	843,894
Hidalgo	569,463	744,258	948,488	1,177,243	1,424,767	1,695,114	1,972,453
Jim Hogg	5,281	5,593	5,985	6,286	6,538	6,468	6,225
Maverick	47,297	55,892	64,984	73,581	81,032	87,850	93,381
Starr	53,597	66,137	79,538	93,338	107,249	120,959	134,115
Webb	193,117	257,647	333,451	418,332	511,710	613,774	721,586
Willacy	20,082	22,519	24,907	27,084	28,835	30,026	30,614
Zapata	12,182	14,025	16,217	18,415	20,486	22,354	23,733
REGION M TOTAL	1,236,246	1,581,207	1,973,188	2,401,223	2,854,613	3,337,618	3,826,001

Table 4- Projected water demands and population growth in Lower Rio Grande Valley.

B. Beneficial Use of Waters - Irrigation District Integration

A fascinating history of the early attempts at Valley irrigation and companies built up around the agriculture industry is contained in an article titled, "Louisiana--Rio Grande Canal Company Irrigation System" at the Texas Historical Commission website. This traces the early development and creation of irrigation companies and districts in the lower Rio Grande Valley. Much of the success of irrigation and later developments can be traced to this single irrigation company. The pump stations, power generating plants, canals and appurtenances are now a part of the Hidalgo County Water Improvement District #2. The Texas Historical Commission discussion:

"Wisconsin native John Closner established the first steam-powered irrigation system in the lower Rio Grande Valley in 1895. Closner successfully grew sugar cane and entered a sample for judging at the St. Louis World's Fair of 1904. The award-winning sugar cane and Closner's promotion of the lower Rio Grande convinced a number of financiers and agriculturists to invest in the region's irrigation and development possibilities. In 1909 H. N. Pharr, J. C. Kelly, John C. Conway, and A. W. Roth formed the Louisiana-Rio Grande (LRG) Canal Company to transform about 40,000 acres of arid land in Hidalgo County into productive farmland. To do this the company built two pumping stations to divert water from the Rio Grande through an elaborate irrigation system to a planned community of small farmsteads. The agricultural success of the LRG Canal Company, its successor the Hidalgo County Water Improvement District #2, and other similar operations in the region resulted in an influx of Anglo farmers and settlers from the mid-western U.S. into this mainly Hispanic region of Texas. The bountiful harvests propelled the Rio Grande Valley to the forefront of Texas agriculture by the mid-20th century and earned the region a reputation as the "winter garden" of Texas."

Figure 9 - Louisiana-Rio Grande Canal Company Irrigation Pump Station (Smokestacks and steam driven pumps).

The TWDB is the state agency charged with collecting and disseminating water-related data, assisting with regional planning and preparing the State Water Plan for the development of the state's water resources, and administering cost-effective financial programs for the construction of water supply, wastewater treatment, flood control and agricultural water conservation projects.

The Lower Rio Grande Valley agriculture industry has limited water supplies due to reduced flow from the Rio Grande and increasing demand from the municipal sector due to rapid population growth. The Rio Grande 2001 Regional Water Plan and the 2002 State Water Plan project significant water shortages for South Texans over the next 50 years. The Lower Rio Grande Valley project, located in Cameron, Hidalgo and Willacy counties, is a ten year study to integrate state-of-the-art irrigation water distribution network control and management and on-farm irrigation technology and management systems in a large-scale demonstration of cost effective technologies that maximize water use efficiency. The project will demonstrate, document, and incorporate the Districts' ongoing conservation projects, and provide coordination between the Districts' staff, agricultural water users, and state and federal technical agencies. The project includes maximizing the efficiency of flood irrigation, demonstrating the effectiveness of all major irrigation technologies and showcasing how to implement the beneficial findings from the field demonstrations to irrigation districts and farmers. The project will assist in the implementation of the agricultural water conservation management strategies identified in the 2001 Rio Grande Regional Water Plan and the 2002 State Water Plan, and will further agricultural water conservation in Texas.

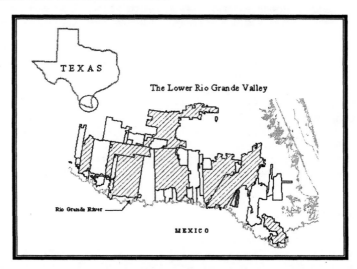

Figure 10 – Eight of the Irrigation Districts which have implemented GIS based water data management systems (Approximately 740,000 acres are irrigated on the US side of the border).

C. Environmental Concerns vs. Development and Border Security

(Newspaper Article- McAllen Monitor 9-7-06) "Age, drought and poor maintenance are part of the problem, but locals blame the U.S. Border Patrol on the worst damage. Agents use the tops of levees as roads for patrols, and they often drag tires behind vehicles to wipe out footprints of illegal immigrants to help detect fresh ones. "The levees were not designed to have that sort of traffic flow on them," Cameron County Judge Gilberto Hinojosa said. The agents are "just doing their job, but it has had an effect on a system that was designed to protect us." Irrigation drains once encased in dirt now have several feet of cement base exposed. In some places, the once-towering levees are almost flush with the surrounding farm land. Mario Villarreal, assistant chief for U.S. Customs and Border Protection's Rio Grande Valley sector, said he was unaware of concerns about agents driving on the levees, since farmers and other law officers do it, too. "Our agents are on the levee; they're on the lower river road. We have a responsibility to protect the nation's border and it is critical that our agents move laterally," he said. Hinojosa told a state Senate Committee on Transportation and Homeland Security in July that the federal government should take responsibility for levee repairs. U.S. Rep. Ruben Hinojosa, a Mercedes Democrat unrelated to the judge, got $6 million added to a House appropriations bill in June to repair the levee, which is about three times more than the annual $2 million allocation for maintenance. But the bill is still pending in Congress. Carlos Marin, acting U.S. commissioner of the federal International Boundary and Water Commission, which oversees the U.S. side of the Rio Grande and its levees, said $6 million is not enough to protect against a flooding disaster. He estimated $125 million was needed for repairs, but South Texas was a low priority post-Hurricane Katrina. "Everybody basically in the country is now concerned about levee systems, so we're competing with other places, like New Orleans," he said. Afraid that **Washington** won't act quickly enough, Hidalgo County officials put a $100 million bond referendum on the November ballot for flood drainage projects. Pete Leal, 61, fears the levee won't withstand another major storm if repairs aren't made soon. In 1967, Hurricane Beulah dumped three feet of rain on the Valley. Flood water surged about 200 yards past the riverbank and lapped the top of the then-sturdy berm. "If anything like that happened now, I'd say it would wipe out Brownsville," Leal said. National Weather Service meteorologist Tim Speece said a storm along the lines of Beulah, which made landfall between Category 3 and Category 4, could hit again. "We've just been fortunate that since 1967 we haven't had anything like that severe flooding," Speece said.

IV. Significant Projects

A. Hidalgo County Protective Levee System –

The USIBWC is considering alternatives to raise the 4.5-mile Hidalgo Protective Levee System to meet current flood control requirements. Alternatives under consideration would raise levee height from 3 to 8 feet, depending on location, and expand the levee footprint by lateral extension of the structure. Levee footprint increases toward the riverside could potentially extend into the Lower Rio Grande Valley National Wildlife Refuge System. Footprint increases toward the levee landside could extend beyond the USIBWC right-of-way. Soil borrow easements would be used to secure levee material. The Project would be implemented in two phases. Phase 1 would raise existing levee height along the 3.3-mile upstream reach of the levee system. Phase 2 would partially re-route the 1.2-mile downstream reach of the levee system to eliminate the need for construction of a floodwall in front of the Hidalgo Historic Pumphouse, a resource included in the National Register of Historic Places. A new levee segment, approximately 0.7 miles in length, would be built along the south margin of the pumphouse intake channel, and the channel would be crossed to tie the new structure to the existing levee system.

B. Development of a Geodatabase –

The Center for Research in Water Resources (CRWR) of the University of Texas at Austin in cooperation with several Mexican agencies is developing a binational dataset to build hydrologic information systems, which can be used to support hydrologic analysis and modeling in the Rio Grande/Bravo basin. This dataset consists of an ArcHydro-based Geographic Information System and relational data base containing hydrologic and hydraulic information from American and Mexican sides.

C. Overall Lower Rio Grande River Levee Improvements –

Overall, the Lower Rio Grande Flood Control Project is estimated to cost approximately $125 M, including environmental documentation, geotechnical investigations, design and construction. Cost for the work has increased from the 2004 work plan due to inclusion of river levees subject to freeboard encroachment, and the interior floodway system. The 2004 work plan addressed only areas where there was levee overtopping. In addition, costs for levee surfacing, environmental mitigation, and inflation have been added and the contingency increased. The costs shown below are for construction only.

1. Hidalgo Loop Levee Phase 1 - 3.3 miles $4.86M
2. Hidalgo Loop Levee Phase 2 - 1.2 miles $5.0M
3. Mission Levee - Peñitas to Main Floodway - 12 miles $23.0 M
4. Common Levee (Anzalduas to Hidalgo Loop Levee) - 5.3 miles $14.0M
5. Lateral A (Hidalgo) to Donna Pump - 11 miles $18.3 M
6. Retamal Dike - 3 miles $5.4 M
7. Donna Pump to Brownsville - 18 miles $19.0 M
8. Interior Floodway Levee Rehabilitation - 58 miles $21.5 M

D. Present Conditions –

Newspaper Article – McAllen Monitor 9-12-06 - "The bi-national body charged with regulating flood controls on both sides of the border said Wednesday it could not prove to the Federal Emergency Management Agency that levees in Hidalgo County could withstand a 100-year flood. The U.S. section of the International Boundary and Water Commission's statement comes one day after FEMA sent a letter asking owners of federal and private levees in the county to prove their levees were structurally sound. "We know from studies that have already been done that (portions of the levees) could be overtopped," said the IBWC's El Paso-based spokeswoman, Sally Spener, on Wednesday. The IBWC was created by a 1944 treaty with Mexico to operate both nations' flood control efforts. A study done in conjunction with the U.S. Army Corps of Engineers in 2003 shows that the river levee on the U.S. side would overflow along 38 river miles from Peñitas to the Santa Ana National Wildlife Refuge south of Alamo.

The Bond Election - Since it seems unlikely the federal government will fund the IBWC, Hidalgo County is stepping in. The county's drainage district plans to place a $100 million bond referendum on the ballot in November. "We need to address our drainage situation as a county community," Garcia said. "…. This is the No. 1 issue in the county." Should the bond pass, the drainage district would supply the IBWC with $10 million for the most critical miles of river levees. The other $90 million would be used for county drainage projects such as the Mission Inlet and the

Raymondville Drain." FEMA telling certain parts of Hidalgo County that it is no longer safe from a 100-year-flood will motivate people to listen and learn about drainage," Garcia said. "A substantial amount of money is needed," he said. "The last time we addressed drainage was in the late 1970s." Since then, the county has lost ranch land to fend off the water....After Beulah, modifications were made and a new floodway that could take in more runoff from the Rio Grande was established, he said. But residential development, which also includes Sharyland Plantation, has since popped up all around the old floodway, which could still be flooded should the Mission Inlet Closure near Bentsen State Park breach. "We can no longer just sit back and watch television every hurricane season and hope it doesn't hit us," Garcia said...."

(Author's Note: The Bond Election did pass in November, 2006 and now the district is making plans to sell the necessary bonds to fund the proposed improvements described above.)

V. References

1. Anderson, Andrew E. 1932. Artifacts of the Rio Grande Delta Region. *Bulletin of the Texas Archeological and Paleontological Society* 4:29-31.

2. Fipps, Guy and Pope, Craig September 1968. "Implementation of a District Management System in the Lower Rio Grande Valley of Texas", Department of Agricultural Engineering, Texas A&M University, College Station, TX 77843-2117

3. Grozier, R.U., Hahl, D.C., Hulme, A.E. and E.E. Schroeder, September, 1968. Texas Water Development Board, Report No. 83, Floods from Hurricane Beulah in South Texas and Northeastern Mexico, September-October 1967,

4. Melden & Hunt, Inc., and Sigler, Winston, Greenwood & Assoc., January, 1975. Master Plan for Storm Water Disposal for Hidalgo County, Texas, Supplemental Report.

5. Melden and Hunt, Inc., and Sigler, Winston, Greenwood and Assoc. 1985. Master Drainage Plan for Mission Inlet and South Rado Drain prepared for Hidalgo County Drainage District No. 1.

6. Melden and Hunt, Inc., 1988. Rado Drain Alternatives - Study for Hidalgo Co Drainage District No. 1.

7. Parsons Corporation- Austin, Texas, September 2005. "Final Environmental Assessment Alternatives for Improved Flood Control of the Hidalgo Protective Levee System" for International Boundary and Water Commission - El Paso, Texas.

8. Patino, Carlos, McKinney, Daene C. and David R. Maidment. "Development of a Hydrologic Geodatabase for the Rio Grande/Bravo Basin," The Center for Research

in Water Resources at the University of Texas at Austin, 10,100 Burnet Road, Austin, TX 78758.

9. L.L. Rodriguez & Associates, Inc. and Bernard Johnson Inc. December, 1983. City of McAllen, McAllen, Texas, Preliminary Engineering Report for the Drainage Improvements to the South Rado Drain, Drainage System.

10. Texas Agriculture Experiment Station, and Texas Water Resources Institute of the Texas A&M University System Texas Water Resources Institute Report: TR-275, September, 2004. "Estimated Benefits of IBWC Rio Grande Flood-Control Projects in the United States," International Boundary and Water Commission, United States Section, El Paso, Texas.

11. U.S. Army Corps of Engineers, Galveston District, September 1968. Report on Hurricane "Beulah," 8-21, September 1967.

12. U.S.D.A.- SCS, July 1969. Comprehensive Study and Plan of Development, Lower Rio Grande Basin, Texas.

13. U.S. Department of Housing and Urban Development, Federal Insurance Administration May 23, 1978. Flood Hazard Boundary Map Hidalgo County, Texas, Unincorporated Area, Community Panel Number 480334 00012-0021.

14. U.S. Department of Housing and Urban Development, Federal Insurance Administration, July 2, 1980. Flood Insurance Study, Hidalgo County Unicorporated Areas, Texas.

15. U.S. Department of Housing and Urban Development, Federal Insurance Administration, December 15, 1980. Flood Insurance Study, City of McAllen, Hidalgo County, Texas.

16. U.S. International Boundary and Water Commission. ERDC TR-03-4, Condition assessment, Report 1, Lower Rio Grande Valley levees, south Texas.

17. U.S. International Boundary and Water Commission. ERDC TR-03-4, Condition assessment Report 2, Lower Rio Grande Valley levees, Texas and New Mexico levees.

18. U.S. International Boundary and Water Commission. ERDC TR-03-4, Condition assessment, Report 3, Lower Rio Grande Valley levees, south Texas–Mercedes to Brownsville.

19. Wyatt, Linda Barton 1967. Beulah Flood Volumes 3 and 4.

THE MAYA: AMERICA'S FIRST WATER RESOURCE ENGINEERS

James A. O'Kon, P.E., M. ASCE

President, O'Kon & Company, Inc., 26-104 Plantation Drive, Atlanta, GA
30324-2959; PH (404) 237-3018; FAX (404) 842-0642

ABSTRACT

The Maya, America's first civil engineers developed unique water resource technologies that successfully supported a dense population of 1800 to 2600 people per
square mile. This unique engineering accomplishment was but one of the technical
advancements achieved by the Maya during their 2000-year history. Archaeologists
considered the Maya to be a "stone age" culture; however, this science-based society
built a scientifically advanced culture while Europe languished in the Dark Ages.
This ingenious culture developed technologies, and sciences that were not "discovered" by the Europeans until the nineteenth century.

These unique technological achievements were the result of integrating native in-
genium with synergistic management to plan and construct a complex infrastructure.
This infrastructure included efficient water management systems that enhanced the
inconstant natural water supply. The Maya engineers developed projects that provided potable water, irrigation of agriculture and aquaculture, water storage systems,
reclamation of storm water, flood control, and road and bridge construction.

These lost landmarks of civil engineering were hidden for centuries by the tropical
rainforest. Recently, the unique accomplishments of the Maya engineers were uncovered by a new field of archaeo-engineering using state-of-the-art forensic engineering
techniques.

THE TECHNOLOGY OF THE MAYA

The technological advancements of the Maya included a legacy of unique scientific achievements. The Maya developed advances in multiple scientific disciplines and
interfaced these scientific advancements to develop unique projects that enabled sur-
vival and enhancement of life for the dense population. These technologies included
disciplines in city planning, civil engineering, architecture, as well as the pure sciences
such as astronomy, higher mathematics, and written language.

The scientific advancements included achievements that were well advanced over European capabilities in the first Millennium. These unique developments included a wide range of scientific achievements including writing, mathematics, astronomy, calendarics, and medicine.

The Maya developed one of the world's five original written languages. They developed hieroglyphic symbols which stood for syllables in the spoken language as well as mathematical characters. The Maya developed a vestigial mathematical system that used a base of twenty and could mathematically calculate 20 to the 20th power. They developed the concept of the number zero 700 years before the concept was introduced to Europeans by Arabic scholars. The Maya developed a calendar that is more accurate than our modern calendar. Their astronomical almanacs included accurate cycles of lunar eclipses, solar eclipses and traced the path of the planet Venus with an error of 14 seconds per year.

To produce their complex infrastructure, the Maya developed techniques for fabricating cement for use in cast-in-place concrete structures 1500 years before Europeans discovered the method for producing Portland cement. Their invention of the "Maya arch" enabled engineers to combine masonry and cast-in-place concrete to create long span structured spaces to create dramatic palaces, temples, and engineering infrastructure. The Maya engineered an infrastructure which included large concrete-lined, underground water reservoirs, arched underground aqueducts, canal systems for efficient and productive agriculture, elevated all weather paved roads, bridges, and canals for transportation and agriculture.

Archaeologists consider the Maya to be a "stone age" culture. However, the technological accomplishments of the Maya have not been fully illuminated by archaeologists. Archaeologists have been more concerned with anthropological issues and have overlooked the engineering achievements of the Maya. They continue to consider the Maya to be "stone age" because, in spite of their obvious scientific and engineering accomplishments, the Maya did not utilize tools of iron or bronze. However, the Maya used tools fabricated from a material that is harder than steel. Their tools were made from jade and jade is a material harder than iron or steel. Jade registers as a seven on the Mohs' scale of hardness while iron and steel are rated as a five (Figure 1).

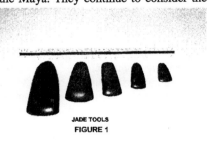

JADE TOOLS
FIGURE 1

THE DEVELOPMENT OF THE MAYA EMPIRE

Centuries before Europeans landed on the shores of the Americas, the Maya developed one of the world's longest enduring civilizations. This society, with advanced levels of technological and scientific capabilities, ruled over an empire that en-

GULF OF
MEXICO

NORTH

FIGURE 2
MAP OF MAYA DOMAIN

compassed over 100.000 square miles, and was governed by at least fifty independent city-states led by warrior-kings who were believed to posses divine powers. This Empire, which had more cities than ancient Egypt, extended from the Yucatan Peninsula in Mexico to the Central American country of El Salvador (Figure 2). The population at the center of the empire is estimated to be 1800 to 2500 people per square mile. This can be compared to modern Los Angeles with a population density of 2300 people per square mile.

Creative Maya engineers shaped the sophisticated infrastructure of this expansive Empire, which reached its peak from AD 250 until AD 900, and produced the first civil engineers of the Americas. Maya engineers constructed gleaming white and polychrome high-rise towers, palaces, pyramids, and temples rising over large, well-planned cities, served by a sophisticated infrastructure that boasted an advanced water management system. This included cast-in-place concrete underground water storage reservoirs, arched subterranean aqueducts, concrete-paved elevated roads constructed above the flooded rainforest floor, extensive canal systems for waterborne transportation and irrigation with raised fields for an abundant and extensive variety of crops to sustain the dense population, and bridges spanning across flooding rivers.

These civil engineers created many unique and ingenious works, most of which have been possessively hidden by the vast and encroaching rainforest. Recently, however, the study of Maya accomplishments has been accelerated by new technologies made available for research. This pursuit was originally the domain of explorers, adventurers, and archaeologists. The adventurers first became aware of the cities of the Maya in 1842 when John Lloyd Stevens published a book describing his discovery of the ruins of an amazing culture. However, today the search has been joined by archaeo-engineers, NASA satellite mapping scientists, and other "cross-over" disciplines. The success of these new disciplines of archaeo-engineers has effectively advanced discovery of the hidden trove of lost the sciences of the Maya using electronic and analytical methodologies. They have uncovered new evidence of the unique works created by the ingenuous Maya engineers. The use of computer simulation and remote sensing techniques have assisted in raising the veil of secrecy created by the dense rainforest. These discoveries are of a nature that is outside the technical capabilities of ground-based archaeologists to identify and interpret.

A UNIQUE METEOROLOGICAL AND GEOLOGICAL ENVIRONMENT

The Yucatan Peninsula which contains the majority of Maya cities has a unique geotectonic composition and a meteorological environment that includes one of the heaviest yearly rainfalls in the Americas. This heavy rainfall occurs during a six-month period, followed by a six-month dry season. The area averages 2000 mm of rainfall per year. However 76% of this rainfall occurs between the months of May to October. The remainder of the year has a paucity of rain and is aptly termed the dry season.

Much of the Maya zone is underlain with a porous limestone stratum that contains an extensive underground aquifer. The heavy precipitation during the rainy season quickly percolates directly into the aquifer. As a result, surface water is scarce despite the heavy tropical precipitation. Few streams or rivers exist in the Maya region and with minor exceptions, no permanent lakes of standing water. The geologic composition of the Maya empire with heavy rains and alternating droughts presented serious challenges for the Maya engineers. They were charged with providing an infrastructure that would enable this unique culture to operate efficiently on a year around basis in order to provide sufficient food and water for the dense populations.

This unique combination of an inconstant meteorological environment and a geotectonic subsurface that drew storm water directly into the aquifer offered unique challenges to the Maya engineers. Practical solutions were required to maintain a constant water supply, provide an adequate food supply, maintain flood control, and construct transportations systems that enabled travel during the flood season.

The goals of the Maya engineers were to solve multiple circumstances attendant to the conservation and storage of water for domestic usage, control of water for agriculture, prevention of flooding, methods for land reclamation, and transportation systems. The projects constructed by the Maya engineers can be categorized into three general areas: (1) Collection and conservation of water for potable water, agriculture, and transportation, (2) Construction of elevated roadways to permit travel during the rainy season, (3) Design and construction of bridges over flooded rivers.

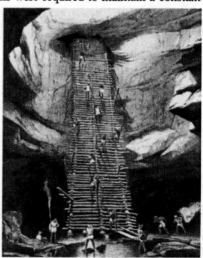

CENOTE: AQUIFER WATER SOURCE   FIGURE 3

MAYA WATER RESOURCES MANAGEMENT

The Maya kingdom includes hundreds of city-states. A fortunate few of these cities were sited near one of the scarce rivers or lakes. Additional cities were sited adjacent to giant sinkholes or "cenote" in the limestone. The sinkholes provided access to the underground rivers and provided a constant supply of water (Figure 3).

The majority of Maya cities were required to rely on Maya engineering creativity to provide a dependable year around source of water. Maya cities preferred sites on high ground which provided a good defensive position. In these locations where a surface water supply was not available Maya engineers relied on several types of water collection and storage systems.

Open Reservoirs

In cities located at sites with natural water drainage features, Maya engineers excavated large in-ground reservoirs to collect water during the rainy season. The reservoirs were located in the path of a natural drainage feature and were excavated into the soil and rock substrate. The base and sides of the reservoir were lined with masonry and covered in plaster to reduce leakage. The reservoirs in each location were sized to satisfy the water requirements of that city. Based on the Maya life style, it is estimated that 2000 gallons per person were required for a 12-month period. The reservoirs were constructed with adjacent water channels that directed water to populated sections of the city. This type of reservoir was used at the Maya city of Tikal, which maintained a system of 13 reservoirs with a total volume of over 39 million gallons of water.

Subterranean Reservoirs (Chultuns)

In Maya cities located on high ground or on level plains, the use of natural drainage from a natural watershed was not possible. In these locations, Maya engineers constructed underground reservoirs with manmade watersheds for collection of storm water. These reservoirs or cisterns are termed "chultuns" by the Maya. The construction of this subterranean water reservoir required excavation into the soil and rock. The interior was sealed by lining the base and sides with masonry and stucco plaster. The top of the chamber was closed with a Maya arch constructed of masonry and cast-in-place concrete. The top terminated in a narrow restricted neck with a small opening for accessing the water reserve (Figure 4).

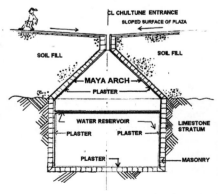

SECTION: UNDERGROUND RESERVOIR (CHULTUNE)
FIGURE 4

To provide a substantial flow of storm water for the chultuns, the Maya located the chultuns under broad paved plazas which were sloped to drain into the chultun (Figure 4). Storm water runoff into the chultuns was provided by these sloped plazas as well as by the sloped roofs of the adjacent structures. Certain large Maya cities, such as Uxmal, contained hundreds of chultuns to provide water supplies for that city. In these cities the construction of these manmade watersheds to supply the chultuns created large plazas that also enhanced urban life.

Maya Water Filters

The state of purity of water stored for six months in a reservoir would present a health hazard. However, Maya engineers developed an ingenious water filter for household use. The filter operates on a similar basis to the modern ceramic filter. The Maya filter device consisted of a one meter tall by 30 cm diameter porous limestone cylinder with a cone shaped lower end (Figure 5). Water retrieved from the storage reservoir was placed into the top of the filter and was permitted to percolate down through the body of the cylinder. After passing through the matrix of the filter, the filtered water was drained into a vessel placed below the bottom cone of the filter.

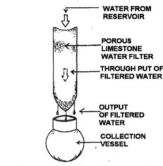

DETAIL: LIMESTONE WATER FILTER

FIGURE 5

Development and Creation of Aqueducts Systems

Maya water management resource technology was uniquely demonstrated in the city of Palenque. This grand city was the exception to other Maya cities that required modification of the surrounding terrain surface or extensive storage facilities for a water supply to extend through the dry season. Palenque is sited below a series of rainforest-covered mountains and there were 56 perennial water springs that flowed swiftly down the mountainsides through natural watercourses that extended randomly across the cityscape.

In Palenque, Maya engineers faced a situation in which the multiple streambeds interfered with the growth of the city. Monumental structures and plazas in the city center could not be constructed without interference from the streambeds. In addition, flooding of the central city occurred during the rainy season as storm water overflowed the streambeds. The engineering solution was to control, collect, and redirect the swiftly flowing water. Maya engineers enclosed and redirected the random water flow within an organized system of underground aqueducts.

Maya engineers redirected the flow of water from the mountain springs and storm water overflow through the construction of an efficient network of nine separate

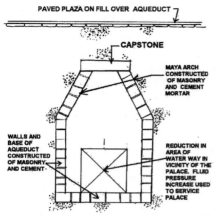

**SECTION: SUBTERRANEAN AQUEDUCT
IN PALENQUE WATER
CONTROL SYSTEM**

FIGURE 6

subterranean aqueducts. This system resolved the flooding issues by redirecting the numerous streambeds into a structured underground system. The aqueduct network was covered with soil thereby creating an additional 23% area of habitable space in the central city. These underground aqueducts were constructed of masonry and concrete. The construction included vertical sidewalls and base and were enclosed at the top with a Maya arch (Figure 6). The majority of this unique system still is operating after 1200 years.

It should be noted that Maya engineers understood the precepts of fluid pressures. Certain aqueducts adjacent to the palace complex were constructed in a manner that reduced their cross section by 50%. This reduction in the swiftly flowing waters within the aqueduct chamber would increase the fluid pressure and enable the water to be transmitted upward via piping into the interior of the palace and other significant buildings.

MAYA WATER MANAGEMENT FOR AGRICULTURE AND AQUACULTURE

Maya engineers employed a series of alternate solutions for irrigation and reclamation of agriculture areas. One of the most creative was the construction of a network of canals that encompassed a grid of raised agriculture fields (Figure 7). This ingenious system was actually a combination of agriculture and aquaculture. The Maya planted corn, beans, tomatoes, squash, fruit trees, and other produce on these raised irrigated fields. The canals were utilized as aquaculture canals to raise fish, mollusks, turtles, algae, water plants, and other aquatic life for use as food.

**SECTION: RAISED FIELDS AND
CANAL SYSTEM**

The canals were periodically cleaned and the detritus was placed on top of the mounded fields. This material was then mixed with crop residue. The result was a nitrogen rich soil which is estimated to have produced

**PLAN: SYSTEM OF RAISED FIELDS
AND CANALS FOR AGRICULTURE**

FIGURE 7

400% more yield than methods used today. This high yield of agriculture was necessary for the dense population in the Maya domain.

ENGINEERING THE MAYA TRANSPORTATION SYSTEMS

The rainy season created a flooded condition in the rainforest for six-months of the year. Muddy trails with streaming water crossing the tortuous rainforest undergrowth made travel very difficult. The economy of the Maya relied heavily on inter-city commerce for both agricultural and other trade goods and for rapid military movements.

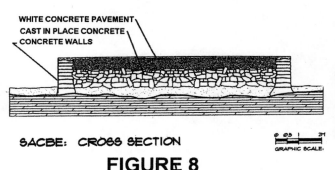

WHITE CONCRETE PAVEMENT
CAST IN PLACE CONCRETE
CONCRETE WALLS

SACBE: CROSS SECTION

FIGURE 8

masonry and cast-in-place concrete.

Maya engineers solved the difficulty of flooding, mud, and rough trails through construction of a network of paved roadways which were raised above the flooded jungle floor. The raised roadway was elevated above the surface using

The surface of the roadway was paved with a white colored cast-in-place concrete paving (Figure 8). The white paved surface of the road led to the Maya name for the roadway. The Maya term these roads "white roads" or "sacbe." These all weather roads permitted travel throughout the year, even at nighttime when the white surface reflected moonlight and starlight.

The system of Maya roadways extended from city to city and to religious centers. Figure 9, indicates some of the routes that have been identified by archaeologists. The 100 km long sacbe between Cobá and Yaxuna has been investigated personally by the author. A portion of this road has been restored, as shown in the photograph taken by the

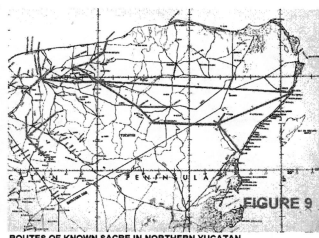

ROUTES OF KNOWN SACBE IN NORTHERN YUCATAN

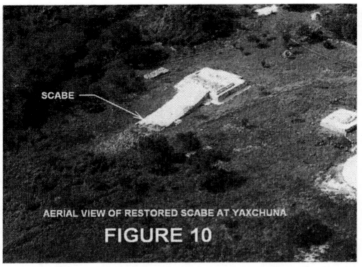

author   (Figure 10).

AERIAL VIEW OF RESTORED SCABE AT YAXCHUNA
FIGURE 10

MAYA BRIDGE CONSTRUCTION

There are numerous short span bridges constructed by the Maya crossing streams. As previously discussed, surface watercourses are scarce and bridge construction was not widely employed. The exception to this was the long span bridge across the Usumacinta River at the city of Yaxchilan on the modern boundary between Mexico and Guatemala. This three-span long suspension bridge, with a 67-meter center span, was the longest bridge in the ancient world (Figure 11).

MAYA SUSPENSION BRIDGE AT YAXCHILAN
FIGURE 11

The city of Yaxchilan was located in a great ox bow in the river and was isolated by 15 meters of floodwater during the rainy season. The ingenious solution developed by the Maya engineers was to construct two tall stone and concrete towers in the river and construct a 106-meter long suspension bridge of hemp rope between the grand plaza of the city and the northern embankment. The bridge platforms were elevated above the 15-meter high floodwaters and provided an all weather access to the city. The high hills inside the oxbow of the river presented a natural fortress for the city and the ingenious bridge structure gave them the access to operate on a year around basis.

The Maya engineers constructed an efficient all weather lifeline that was a three span tension structure founded on masonry and cast-in-place concrete piers. The Maya engineers understood the importance of hydraulic shapes. The bridge piers are "D" shaped in plan with the curved face extending upriver into the oncoming water flow (Figure 12). The knowledge of scour was an important factor in this bridge design. The piers extended down to bedrock. Large flat stones were connected together to form the vase and the concrete structure was constructed upon the flat stone. These flat stones are in place today after over 1300 years of water flow.

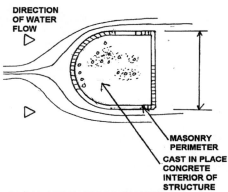

PLAN: HYDRAULIC SHAPE
OF BRIDGE PIER
FIGURE 12

REFERENCES

FITZSIMONS, N. (1995). "The Greatest Bridge Spans in the World from Earliest Times to the Present"

HARLOW, G. (1991). "Hard Rock," *Natural History Magazine*, August

MAUDSLEY, A. P. (1889). *Biologia Centrali-Americana,* Francis Robicsek

O'KON, J., et al. (1989). *Guidelines for Failure Investigation*, ASCE

O'KON, J. (1995). "Bridge to the Past," Civil Engineering Magazine, January

SCARBOROUGH, V. L. (1999). *Economic Aspects of Water Management in Prehistoric New World,* JAI Press

SCHELE, L., and MILLER, M. E. (1990).*The Blood of Kings*, George Braziller

SHARER, R. J. (1994). *The Ancient Maya*, Stanford University Press

STEPHENS, J. (1843). *Incidents of Travel in Yucatan*, Harper & Brothers

TATE, C. (1993). *Yaxchilan, the Design of a Mayan Ceremonial City*, University of Texas Press

The *Acequias* of San Antonio and
the Beginnings of a Modern Water System
Dana Nichols

Abstract

In 1718 Spanish Missionaries established permanent settlements in San Antonio
and immediately began constructing a system of irrigation ditches, or *acequias*,
to divert water from the San Antonio River and San Pedro Creek to irrigate
hundreds of acres of farmland. The *acequias* were so well built that the Acequia
Madre, built between 1731 and 1745 still provides irrigation water for 400 acres
of farms, establishing it as the oldest continuously working *acequia* in the
country. This presentation will follow 300 years of water infrastructure in San
Antonio and along the way touch on how important water is to a thriving
community.

A City Born of Water
For the last century or so San Antonio, Texas, has relied upon the artesian wells of
Edwards Aquifer for its water supply. The Edwards feeds the San Pedro and San
Antonio springs which, until the middle of the 20th century, provided the base flow
for San Pedro Creek and the San Antonio River.

In 1716 Spain and France were at war and the fighting carried over to the New
World. The French controlled most of the gulf coast, including the mouth of the
Mississippi River. Spain saw this as a threat to their holdings in New Spain and
developed a three-pronged approach to securing their New Spain borders. The plan
was unique in that rather than conquering and displacing the native population, the
plan called for the natives to become the defenders of New Spain. Missions would be
established to gather the natives to be converted to Christianity and missionaries
would teach the principles of farming in preparation for establishing new colonies.
Second, soldiers would establish a *presidio* (army post) near the mission to keep
order. Finally, once native farmers had established settlements, Spanish colonists
would come to these new settlements to begin new lives. In 1716 a Spanish *entrada*
(expedition) was sent to east Texas to secure the French border and six missions and

Conservation Manager- Outdoor Programs; San Antonio Water System; 2800 Hwy
281 North; San Antonio, Texas, 78212;dana.nichols@saws.org; 210-233-3656

a *presidio* were established. The plan was working, but leaders shortly realized that they needed to establish a settlement half way between the border and the capital of New Spain, Mexico City.

Previously identified by Spanish missionaries as a desirable location for a settlement, Father Olivares and Don Martin de Alacon established Mission San Antonio de Valero (The Alamo) between the San Antonio River and San Pedro Creek in 1718. By 1731 there were six missions and a *presidio* in what today is the City of San Antonio. Early Spanish settlers knew that the successes of settlements were dependent on the success of farming and that in this semi-arid region this meant irrigated fields.

Acequias
The Spanish were well versed in the construction and use of large-scale irrigation systems, and these irrigation systems, or *acequias*, had been widely used in Spain since the Moorish conquest.
The primary water distribution system in the area was the *acequias*, or community water ditches. This extensive network of irrigation canals began with the first ditch, Pajalache or Concepcíon Ditch, which became operational about 1720 and remained in operation until 1869 when it was abandoned. The San Francisco de la Espada Mission *acequia* was built between 1731 and 1745 and remains in working order today. The *acequias* were supplemented by shallow wells and provided water for both irrigation and consumption.

In addition to the irrigation canals built by the missions, several public *acequias* and ancillary ditches were built in the San Antonio area to serve the *presidio* and the lands of the Canary Islanders at the Villa de San Fernando. Located in present-day downtown San Antonio, today this area is known as La Villita. The construction and maintenance of the *acequias* required considerable amounts of labor, and some of the larger canals took more than two decades to finish. Because of the relatively high cost of the *acequias*, the amount of irrigated land was limited, and competition for such land was strong. Much of the better land went to the Canary Islanders, who constituted the local political and socioeconomic elite.

Water rights were strictly controlled and were sometimes sold or bought separately from the land. Landowners were expected to help dig new irrigation ditches and to defray the expense of upkeep. Those who failed to comply with regulations to keep the canals in working order were subject to fines.

After the missions were secularized in the early 1790s, the city authorities undertook to oversee the distribution of water. City control was discontinued in the later half of the nineteenth century, and the remaining *acequias* were operated for a time as informal community enterprises or, in the case of the San Juan *acequia*, by an incorporated mutual company.

These ditches also began to serve as a *de facto* sewer system. Early San Antonians merely deposited their garbage and other wastes into the ditches where they were carried downstream. In 1836, the San Pedro Ditch was reserved for drinking and cooking water only; penalties were established for using it for bathing or as a sewer. Although crude, this water and wastewater operation served the City's needs until 1866 when a severe cholera epidemic prompted real efforts to establish a satisfactory water supply system.

Surface Water to Ground Water - Brackenridge System

Many water development proposals were discussed and subsequently discarded over the years until the City finally entered into a water supply contract with J.B. La Coste and Associates on April 3, 1877. La Coste constructed a pump house near the headwaters of the San Antonio River in what is now Brackenridge Park. Water pressure operated a pump that lifted water to a reservoir near the old Austin highway on the present site of the Botanical Garden. This site was high enough for the water to flow by gravity into the distribution system.

In 1883 a new company, led by George W. Brackenridge, acquired the water system. Recognizing that the source of the springs was possibly a subterranean reservoir under high pressure, Brackenridge proposed that his firm purchase property along the river and drill a well. In 1889, the first artesian well was bored in what later became Brackenridge Park. Two years later an eight-inch discovery well was drilled to a depth of 890 feet at Market Street and the San Antonio River. By 1900, all of the system's water was obtained from artesian wells linked directly to the distribution system.

In 1905, George Brackenridge sold his interests in the water company to George Kobusch of St. Louis. At that time the name was changed to the San Antonio Water Supply Company. Shortly thereafter, Mr. Kobusch sold the business to a Belgian syndicate. While it was under foreign ownership, the water company was known as "Compagnie des Eaux de San Antonio" and was managed by the Mississippi Valley Trust Company of St. Louis, Missouri.

The City Takes Ownership

Partly to recover some of their financial losses from World War I, the Belgians sold the waterworks to a group of local investors in 1920. Contract and rate disagreements marred the relationship between the City and the new water entity. In 1924, the company demanded a rate increase, and since an agreement could not be reached, the new rates were put into effect and the City was enjoined from interfering. This situation prompted the City to issue seven million dollars in revenue bonds and purchase the system outright. On June 1, 1925, the utility became known as the City Water Board and its management was placed under the Board of Trustees appointed by the San Antonio City Council. At the time of the purchase pumping was an average of 25 million gallons a day serving 38,000 people.

During the Depression and the war years the City Water Board was able to keep pace with increasing demand without much difficulty. However, the post-war building boom and the impact of the 1950s drought significantly taxed the Board's capabilities. In the mid1950s the water operation utilized many widely scattered secondary pumping stations which were designed to serve immediately adjacent neighborhoods. These stations essentially operated independently and did not provide adequate system redundancy.

Water Planning – 1954-1991

A master plan for improvements was approved in 1954. During the 1960s, 1970s, and 1980s both the water and wastewater systems continued to expand as customer demand increased. In 1965, the City built a new wastewater treatment plant, and throughout much of this period the City Water Board was involved in negotiations or court actions involving attempts to secure a supplemental water supply. In 1979, a committee established by the City Planning Commission reported to the City Council that San Antonio should pursue the necessary federal and state permits to construct San Antonio's first surface water supply project known as the Applewhite Reservoir. Shortly thereafter, the Council passed a resolution directing the City Water Board to initiate the permitting process. The City Water Board received the necessary permits from the Texas Water Commission in 1982, and the U.S. Army Corps of Engineers in 1989. Construction on the Lake began a few months later.

On May 4, 1991, the citizens of San Antonio, by a narrow margin, voted to discontinue the Applewhite Project. In the following months the Board of Trustees of the City Water Board voted to sue the City over the legality of the election. Court action subsequently upheld the City's position and Applewhite construction was halted.

While water issues garnered the most attention, wastewater continued to be a demanding subject. During the 1970s and 1980s the City continued to upgrade its waste water facilities. The oldest treatment facility which had utilized Mitchell Lake for sludge disposal was closed and the lake, a long known favorite bird watching location, also known for its distinctive odor to a generation of San Antonians, was no longer used for treatment. Mitchell Lake was declared by the City Council as a bird refuge in 1987. Today it is owned by the San Antonio Water System and is on long-term lease to the National Audubon Society. It is presently known as the Mitchell Lake Audubon Center.

In 1989 the City of San Antonio asked the State Legislature to create a district devoted to reuse of the municipality's effluent. The bill was signed by the Governor on June 16, 1989. In 1991, the District applied for a permit to divert water from the Leon Creek Wastewater Treatment Plant for reuse purposes. The City Water Board opposed that action due to its possible impact on the Applewhite permit.

San Antonio Water System (SAWS)
The controversy brought on by competing water agencies as well as the Applewhite challenge prompted the City Council to vote in December 1991 to dissolve the 66-year-old water utility and establish a single utility responsible for water, wastewater, stormwater, and reuse. The refinancing of $635 million in water and wastewater bonds made the merger possible. A new entity, The San Antonio Water System (SAWS) became a reality on May 19, 1992.

Sources:

Handbook of Texas Online, s.v. ","
http://www.tsha.utexas.edu/handbook/online/articles/AA/ruasg.html (accessed January 9, 2007).

H₂O University – Tools for Educators; San Antonio Water System Education Department;
http://www.saws.org/education/h2o_university

THE MACHINE OF MARLY
WATER SUPPLY FOR VERSAILLES

Georges Comair

Junior Student, University of Houston, Dept. of Civil & Environmental Engineering, Houston, TX 77204-4003 gfcomair@uh.edu

Abstract
The city of Versailles sits 130 to 140 meters above sea-level, surrounded by wooded hills. In 1661, Louis XIV decided to construct his palace in Versailles. The main problem that was facing the engineers was to supply water to the king's palace and its gardens and fountains. The king wanted the water to be furnished continuously. Due to the palace location, supplying water to the site was like a dream and even the several links between the nearby ponds and reservoirs were becoming inefficient. Two brilliant engineers "Arnold de Ville" and "Rennequin Sualem" started the construction of the "Machine de Marly" that was considered the biggest hydraulic system ever built. This machine was able to deliver water from the Seine River at an elevation of 162 meters and pumped nearly five thousand cubic meters of water a day. This gigantic machine solved the problem for several years until the 19th century when it was destroyed because of inefficiency.

Historical background
In 1661, Louis XIV acceded to the throne of France and got as a heritage from his father the late King Louis XIII, the Versailles Domain, which consisted only of a big swampy forest with a very modest mansion constructed with bricks and stones and used as a "hunting relay" for post horses.

Instead of living at the Louvre Palace in Paris as his father did before, King Louis XIV decided to settle down in Versailles, far from the rebellious population of the capital. For the splendor of his region, he started building a gigantic palace "The Chateau de Versailles" aiming to be considered as the Symbol of his Kingdom. The design of the Chateau included several interesting features among them very large irrigated landscape gardens.

Besides the importance of the architecture, the key feature of the Palace was the hydraulic planning which includes very large irrigated landscape gardens supplied by basins with water delivered through fourteen hundred fountains containing 1400 water jets linked together by a 30 km pipe network. This hydraulic system constituted a nightmare for the chief architect of the kingdom since the Versailles neighborhood was very scarce with water resources and the water consumption for the landscaping would be above 7000 m^3 per hour. The main difficulty facing the engineers was how were they going to supply water for the gigantic construction and the fourteen hundred fountains? (One solution is in Figure 1.)

Figure 1 Overview of the Proposed Machine of Marly (Source: Pendery 2000)

Figure 2 Map Showing the Versailles Hydraulic Projects (Sources: Monnier 2003)

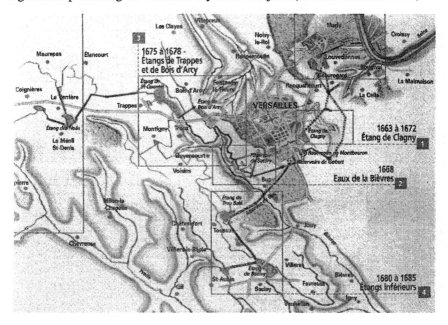

Geographical Location

The swampy forest of Versailles is located about 17 km southwest from the center of Paris. The palace is located at a high elevation, whereas most of all the ponds and basins were lower. This particular characteristic made it nearly impossible to convey water to the construction site using gravity.

The engineers, at first priority, thought about conveying the water from Clagny pond and then from the water course of the Bievres using manual pumps activated by horses. These sources (Clagny and Bievres) could not fulfill the need of the gardens and the fountains. (Figure 2)

Actually, the location of the palace of Versailles was poor when it comes to water supply. The chief engineer responsible of the fountains for the Versailles palace summarized in a clear way the tough situation by saying: "The king couldn't choose a worse place for building his palace."

Engineering Layout

Engineers Gobert and Picard established a chain of canals and reservoirs as seen in Figure 2 to be able to supply the fountains of Versailles with water without interruption. Although a considerable quantity was already available from sources and network of reservoirs, it was nearly impossible for the engineers to catch all the water they needed from lands located above Versailles. This network of canals, basins and reservoirs was depending a lot on the water precipitation and the running water was not potable. Therefore, the idea of basins and reservoir was linked with the hydrology of the region and the rate the water would precipitate.

In 1680, the king took the decision of inviting engineers throughout the country to submit their ideas and inventions in order to solve the problem. One person who had already dealt with the same kind of issues was Arnold de Ville and his friend: a famous contractor named Rennequin Sualem. These two engineers had already installed a gigantic pump in their own estate at Modaves. In 1681, Arnold de Ville and Rennequin Sualem started the construction of the "Machine de Marly" which was at the time considered a wonder of the world. In fact, the machine was the biggest hydraulic system ever built. It was located at the bottom of the hill of Louveciennes on the banks of the Seine in the Marly area, about 12 km from Paris.

Structural Details

The enormous structure on the river needed 1800 construction workers and it took about 800 tons of lead and 17000 tons of steel and 100,000 tons of wood. The machine was designed to produce as much as five thousand cubic meters of water a day.

A small dam concentrated the water flow down to fourteen gigantic waterwheels each about 36 feet (12m) in diameter. These waterwheels supplied two hundred and twenty-one pumps out of which 64 pumped water from the Seine. (Figure 3&4) This was done at three levels:

First level:
The water was pumped to the Mi-Côte reservoir, at 50 meters high above the Seine River.

Second level:
From the reservoir, the water was pumped again at 56 meters high to another reservoir.

Third level:
From the second level to another 56 meters high reservoir, this was linked to the Louveciennes aqueduct.

The water was pumped 162 meters high along a path of 1 km. This was a marvelous achievement for the two experts in hydraulics: Arnold de Ville and Rennequin Sualem.

Designed to deliver 6000 cubic meters of water per day, the machine worked at 700 Horse power, and cost 3,700,000 livres which is about $72,150,000.

The several pieces of the machine were undergoing great stress and strain during operation and a constant supply of spare parts were needed. Soon, the machine efficiency began to decrease and heavy annual bills for maintenance were paid (about $1,050,000). Therefore, after 133 years of good service, the "Machine of Marly" built under Louis the XIV was destroyed. At that time it was pumping only 200 cubic meters of water per day!

After the construction of the Marly machine, King Louis XIV decided to divert the waters of the Eure River. The work started by the construction of an 80 km channel in order to link Pontgouin at the Eure River to the La Tour water pond at Rambouillet. This infrastructure includes a 17 km aqueduct build on three stages using an arch structure. The entire project was supposed to supply the Versailles palace with 50,000 cubic meters of water per day and solve the problem. Delayed many times, it never was completed.

Figure 3 The Marly Machine (Source: Descamps and Monnier 2003)

Figure 4 Waterwheel Mechanisms (Source: Descamps and Monnier 2003)

Conclusions

The Marly machine was still used until 1802 when Napoleon decided to destroy it. Several versions of the machine were constructed year after year. The famous engineer Dufrayer constructed an innovative machine that lasted from 1859 to 1963; this machine was using new hydraulic techniques and steam engines to supply potable water to the Versailles populations. The last solution undertaken was to deliver water to its 300,000 inhabitants using electric turbines and deep wells.

Some people would criticize the machine and call it a "monster", and also some French politicians would criticize the great amount of money used to construct it. Nevertheless, everyone would agree that this "monster" was really at the time, a great achievement in water resources engineering.

References

Barbet (1907). *"Grandes Eaux a Versailles."*

Belidor, Bernard (1784). *"Architecture hydraulique, ou L'art de conduire, d'élever et de ménager les eaux pour les différens besoins de la vie."* Bibliotheque nationale de France. http://visualiseur.bnf.fr/CadresFenetre?O=NUMM-85683&M=notice&Y=Image

Descamps, P. (2003). *"D'exorbitants besoin en eau."* Les cahiers de Science et Vie, April 2003, No. 74, pp. 84-89.

Lay, M.& J. (1998). *"La Machine de Marly."* Exposition booklet.

Lobgeois, P. & Givry, J. (2000). *"Versailles, Les Grandes Eaux."* JDG publications.

Monnier, E. (2003). "Eau: deux projets trop grandiose." Les cahiers de Science et Vie, April 2003, No. 74, pp. 90-97.

Pendery, D. (2000) "La machine de Marly." http://world.std.com/~hmfh/machine1.htm

The Florida Water Management History Project

Garrett Wallace,[1] Buddy Blain[2] and Kathryn Mennella[3]

This paper/presentation was sponsored by the Five Water Management Districts in Florida and the Water Management Institute.

[1] Director- Legislation and Operations, South Florida Water Management District, 3301 Gun Club Road, West Palm Beach, FL 33406; PH (561) 662-7208 email: gwallace@sfwmd.gov
[2] Attorney-at-Law, 801 S. Boulevard, Tampa, FL 33606 PH (813) 253-0242
[3] General Counsel Office, St. Johns Water Management District, 4049 Reid Street, Palatka, FL 32177; PH (386)329-4215; email: kmennella@sjwmd.gov

Abstract

Florida's five water management districts have joined forces with the Water Management Institute (a private not-for-profit 501(3) (C) educational corporation) to capture and chronicle the History of Water Management in Florida. The project has been split into two major components: undertaking a series of oral history interviews with people that have been important to the History of Water Management in Florida; and developing an archive for documents and other information. Florida will have the 50th Anniversary of the 1957 Water Resources Act in 2007.

The Oral History Component

The objective of the oral history component is to provide an oral history of the evolution of water management in Florida from 1947 to the present. It will address how and why Florida's unique system of water management began and has changed, addressing both water management policy and the institutional and physical nature of Florida's water management delivery system. An initial list of 50-60 individuals that have had significant influence on the history of water management in Florida was developed. Subsequently, the list was greatly expanded to ensure that key individuals representing all regions of the state were included. The Samuel Proctor Oral History Program at the University of Florida was engaged to conduct interviews, tape and transcribe the interviews and manage the recordings and transcripts. Depending on the individual, an interview may last for a few hours or may be more extensive requiring repeat interviews.   Each of the interviews will be conducted by a professional skilled in the art of oral history and then transcribed.

The Archive Component

Many documents related to the History of Water Management in Florida are managed in perpetuity as part of the public record by the water management districts and other agencies, but many documents are not. Some materials in the public record have record retention plans and are actually discarded after some period of time. Private collections often contain interesting and unique information that is simply not available in the public record and all too often this information is discarded and lost. The Project seeks to collect this information and materials, particularly the information and materials in private collections.   We have contracted with the University of Florida, Levin College of Law, Legal Technology Institute to develop a professional archive of information and materials.   The archive component of the project will develop a comprehensive index so that materials, once archived, can be quickly retrieved.   Additionally, an electronic archive will be designed so that key materials can be retrieved via the Internet. The oral histories will be available on the electronic archive. A physical archive of original materials will also be maintained.

Highlights in the History of Florida Water Management

1868 — Florida's first water pollution law established a penalty for defiling or corrupting springs and water supplies.

1881 — The Board of Trustees of the State Internal Improvement Trust Fund conveyed 4 million acres in central and southern Florida to Hamilton Disston to drain to attract agriculture.

1893 — The earliest Florida law on drainage of swamp and overflowed lands was enacted.

1905 — The Board of Drainage Commissioners was created to oversee drainage in the state.

1949 — The Legislature created the Central and Southern Florida Flood Control District, which became known as the South Florida Water Management District. Its primary responsibility was flood control, with other duties later added to include water supply management and water conservation.

1955 — The Water Resources Study Commission presented a report to Gov. Leroy Collins that resulted in passage of the Water Resources Act of 1957, which, for the first time, set a water policy in the state. (Florida will have the 50th Anniversary of the 1957 Water Resources Act in 2007.)

1961 — The Legislature created the Southwest Florida Water Management District.

1968 — Florida's new constitution made the protection of natural resources a state priority.

1972 — "The Year of the Environment" sees passages of the Florida Water Resources Act, which created the water management districts as we know them today; the Land Conservation Act, authorizing the sale of bonds to buy endangered lands; the Environmental Land and Water Management Act, creating development of regional impact and area of critical concern programs; the Comprehensive Planning Act, requiring a state comprehensive plan; and the federal Clean Water Act, setting "swimmable and fishable" goals for all U.S. waters.

1973 — Governing Board members were appointed to water management districts statewide.

1976 — Voters authorized water management districts to levy ad valorem taxes, and the Legislature placed consumptive water use permitting under the exclusive authority of water management districts.

1981 — Save Our Rivers program enacted by the Legislature.

1982 — The constitutionality of water management district taxing powers is affirmed by the Florida Supreme Court.

1987 — Surface Water Improvement and Management Act is enacted.

1988–1990 — During a review by the Environmental Efficiency Study Commission of the water management districts, the Legislature conducted a sunset/sundown review and reenacted Chapter 373, providing the districts with new provisions for financial and program performance accountability.

1990 — Preservation 2000 Act enacted by the Legislature, to preserve Florida's imperiled lands and waters.

1994 — Environmental resource permitting consolidated.

1996 — The Water Management District Review Commission made 80 recommendations and, ultimately, considered and approved a comprehensive legislative package. Among the requirements, the water management districts were required to submit priority lists and schedules for establishing minimum flows and levels.

1997 — The Legislature defined regional water supply planning responsibilities of the five water management districts.

1999 — Florida Forever program enacted by the Legislature.

Summary

The Florida Water Management History Project has two major components: undertaking a series of oral history interviews with people that have been important to the History of Water Management in Florida; and developing an archive for documents and other information. In 2007, Florida will have the 50th Anniversary of the 1957 Water Resources Act (1957- 2007).

History of the USDA-ARS Experimental Watersheds on the Washita River, Oklahoma

Jurgen D. Garbrecht, Patrick J. Starks, and Jean L. Steiner

Grazinglands Research Laboratory, USDA, Agricultural Research Service, 7207 West Cheyenne Street, El Reno, Oklahoma, 73036; PH (405) 262-5291

Introduction

Establishment of a national experimental watershed program grew out of early, depression era efforts by the Civilian Conservation Corps and the Soil Conservation Service (SCS) with the 1930's conservation motto of "stop the water where it falls". A central component of this watershed program was the Agricultural Research Service's (ARS) National Experimental Watershed Network that was authorized by Senate Document 59 (Browning et al., 1959).

Three large watersheds of this network are on the Washita River in Oklahoma. They are the Southern Great Plains Research Watershed (SGPRW), which was in operation from 1961 to 1978), the Little Washita River Experimental Watershed (LWREW), which started in 1961 and is still active today, and the Fort Cobb Reservoir Experimental Watershed (FCREW), which was added to the ARS watershed network in 2004 and is still operational. These experimental watersheds have become increasingly important research sites for multi-agency, multi-disciplinary research involving flood reduction, soil and water conservation, agricultural water quality, sedimentation, erosion, energy fluxes, soil moisture, and remote sensing. The SGPRW, LWREW and FCREW are located in south-central Oklahoma, approximately 40 to 50 miles south-west of Oklahoma City. The SGPRW consisted of 13 adjacent watersheds tributary to the Washita River, and covered an area of about 1,130 square miles (Figure 1). The LWREW was part of the SGPRW, but the FCREW was not. In 1978, 12 of the watersheds in the SGPRW were closed, and only the LWREW, with a drainage area of 236 square miles, was retained for research. The FCREW drains an area of 304 square miles and is situated approximately 30 miles north-west of the LWREW (Figure 1).

Prior to farm settlement, which began around 1900, timber and range grasses covered the watersheds. Major land resource areas in the region are the Central Rolling Red Plains, Central Rolling Red Prairies, and Cross Timbers and Reddish Prairies. From a hydrologic standpoint, this 19th century cover was excellent (Allen and Naney, 1991). After the land was settled, the native vegetation was progressively converted to cultivated land. High prices for wheat and cotton during World War I (1914-1918) lead to a great increase in cropped land. The amount of land in cultivation increased until the mid-thirties, when the great depression and dust bowl conditions caused an exodus of farmers from Oklahoma. As the amount of upland cultivation increased, so did soil erosion rates. Annual upland erosion rates probably increased from a fraction of a ton per acre before settlement to 15 to 20 tons per acre during the peak cultivation period. During the 1940s and especially the 1950s, a large portion of the upland cultivation was discontinued, and most of this land was retired, sprigged to Bermuda or seeded to Love grass, and allowed to naturally restore itself to rangeland. Over the last four decades, land use has not changed, except for some reduction in row crops and an increase in small grain production.

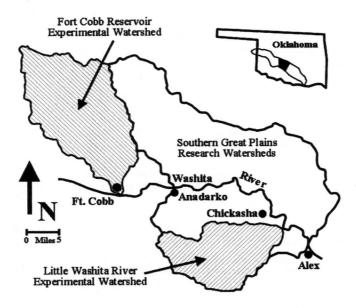

Figure 1. Location of the Southern Great Plains Research Watershed, the Little Washita River Experimental Watershed, and the Fort Cobb Reservoir Experimental Watershed in central Oklahoma.

Since 1961, the USDA-ARS experimental research watersheds in Oklahoma have been used to address issues and problems relating to flood control, soil conservation, watershed hydrology, and water quality. In the following, a short description of watershed instrumentation is given and major research conducted on these watersheds is summarized. Much of the early history presented herein was compiled from communications with retired researchers that have worked on these watersheds and from numerous reports and publications, most notably the report by Staff, Water Quality and Watershed Research Laboratory (1983), and Allen and Naney (1991).

Early Soil and Water Conservation Practices and Research and Demonstration Activities

Conservation work on the watersheds began in 1936 when the Civilian Conservation Corps demonstrated soil-erosion control practices. Erosion control practices included terracing, drop-structures, gully plugging and tree planting. However, it was the long history of flooding along the Washita River that drew attention to this area of Oklahoma. Floods were devastating and caused millions of dollars of damage and loss of life and property.

In 1944, under the umbrella of the Flood Control Act, the Soil Conservation Service (SCS) began efforts to reduce flooding, and conserve soil and water resources in the Washita River Basin. Beginning in 1946, the SCS applied extensive soil and water conservation measures that included terraces, diversions, farm ponds, flood-water retarding structures, gully plugging and smoothing, scrub timber removal and land use planning.

In 1954, the Watershed Protection and Flood Prevention Act (Public Law 83-566) was enacted and broadened activities on the Washita River Basin that were begun under the Flood Control Act of 1944. Conservation projects under Public Law 83-566 included watershed protection, flood prevention, and agricultural and non-agricultural water management. Between the years 1954 and 1985, over 150 flood-water retarding structures were built by the SCS on the SGPRW, with 45 on the LWREW.

In September 1959 a report was presented to the United States Congress titled "Facility Need – Soil and Water Conservation Research" popularly referred to as Senate Document 59 (Browning et al., 1959). As a result of this document, research on the Washita River was initiated in 1960 by the ARS of the U.S. Dept. of Agriculture (USDA). As recommended by this document, the USDA-ARS in 1961 established the Southern Great Plains Research Watershed (SGPRW). Research offices were located in Chickasha, Oklahoma.

The Southern Great Plains Research Watershed from 1961 through 2005

Southern Great Plains Research Watershed (1961-1978)

In 1961, the Southern Plains Hydrology Research Center was established in Chickasha, Oklahoma. The mission of the Center was to assess the overall effects of the USDA-SCS flood control and watershed protection program on the Washita River Basin for the purpose of improving similar programs in the future. The research was to be conducted on the Washita River and major tributaries within the SGPRW (Figure 2).

Installation of flood-retarding structures by the SCS in the SGPRW begun in 1958. One hundred thirty eight flood-retarding structures were completed by January 1979 and controlled a little over 30% of the total drainage area. Of these flood retarding structures 39 were on the LWREW. Daily precipitation data was collected (starting in May 1961) by a basic network of 168 rain gauges that were arranged on a 3 by 3 mile grid. Additional rain gauges were added over time until a maximum of 230 gauges were in operation in 1972. Additional weather variables were measured at two climate stations. Continuous runoff, stream flow and sediment data were collected at 15 stream gauging stations located on the 11 major watersheds of the Washita River Study Reach, and at 6 stations on the Washita River (Figure 2). In addition, 22 small unit-source watersheds were operated to estimate runoff and erosion from small, single land-use areas, and 88 groundwater observation wells were installed to monitor groundwater levels.

Figure 2: The Washita River Reach Watersheds of the SGPRW (1961-1978); tributaries and stream gauging locations. (adapted from Staff, 1983).

In the early years, research addressed issues of rainfall, runoff and sediment measurement techniques and evaluation procedures. Research regarding flood abatement and conservation impacts followed soon afterward. Major results included the identification of changes in watershed runoff characteristics and reduction in sediment yield as a result of flood retarding structures. Runoff volume was not affected by the flood retarding structures, but peak flow rates and overbank floods were reduced, middle and low flow rates were increased, and hydrograph recession was lengthened. Sediment yield was sharply reduced immediately below flood retarding reservoirs in the tributary watersheds, but no corresponding evidence of sediment yield reduction was found downstream on the Washita River. Starting in 1970, attention was also given to water quality issues in the SGPRW. Examples include increased salinity in impoundments and in seepage flow below earthen dams; flow paths of agricultural chemicals in watersheds; and nutrient and sediment yields from agricultural watersheds. Research was also conducted on characterizing hydrologic watershed variables by use of remote sensing techniques, and the rain gauge network supported investigations of stochastic weather generation methods and development of the weather generator CLIGEN.

Little Washita River Experimental Watershed (1978-1985)
In 1978, the Washita River Study Reach was scaled back, and the original mission of the Southern Great Plains Hydrology Research Center was refocused on water quality issues. From the original SGPRW, only the LWREW was retained for further research activities. This watershed was selected as one of seven areas in the United States for the Model Implementation Project (MIP), which was jointly sponsored by the USDA and the Environmental Protection Agency (EPA). The main objective of the MIP was to demonstrate the effects of intensive land conservation treatments and control of non-point sources of pollution on water quality using Best Management Practices (BMPs) in watersheds that are larger than about 25 square miles.

During this MIP investigation period, the rain-gauge network in the LWREW consisted of 36 rain gauges of which one was co-located at an ARS weather station (Figure 3). Flow discharge, suspended sediment, sediment size distribution and water quality parameters were measured at two stream gauging stations, and 24 groundwater observation wells were operated to monitor water quality and

Figure 3. The Little Washita River Experimental Watershed (1978-85); rain gauge network and locations of stream gauging stations and unit source area watersheds (adapted from Allen and Naney, 1991).

groundwater levels. Also, 11 unit source area watersheds were instrumented to quantify runoff and water quality of small streams that discharged into the Little Washita River. The MIP study was completed in 1985.

The research showed that while these conservation practices dramatically reduced sediment and phosphorous losses from upland areas and small subwatersheds, their effectiveness at reducing watershed scale exports depends to a large extent on their level of coverage and placement in the watershed. Unit source and subwatershed scale response to BMPs were not being translated to significant reductions in nutrient exports from the LWREW. This was at first perceived to be a failure of the BMPs and to a lesser degree the MIP. However, what was eventually realized from this research has become the cornerstone of current watershed management planning and remediation both nationally and internationally. First, while there is a minimum level of BMP coverage needed within a watershed to control soil erosion and nutrient movement, broadly applied BMPs over a large area will not affect a reduction in sediment and nutrient export unless high sources of sediment and nutrient export are targeted for specific remediation. Based on these findings, much research has been transferred into successful field tools that identify the major hydrologic, chemical and land management factors controlling critical sources and transport of sediment and nutrient within watersheds. Second, there is a time-lag between BMP implementation and measured reductions in sediment and nutrient exports. Large amounts of nutrients are stored in the soil and sediments of the fluvial system. These can be released slowly, even if all upland erosion and nutrients inputs are stopped. These are important considerations when assessing effectiveness of BMPs and budgeting for long-term observations.

Little Washita River Experimental Watershed, 1985-1992
In 1985 at the end of the MIP, the monitoring activity on the LWREW was further scaled back. The rain-gauge network was reduced to 14 rain gauges. The two stream gauging stations on the Little Washita River were discontinued. All but two unit source area watersheds near the mouth of the LWREW were discontinued. These two active unit source area watersheds were used to monitor channel instability, gullies and their remediation, and were discontinued in 1989. All groundwater monitoring ceased and only one new stream gauging station with a drainage area of 3.4 square miles was operated on Chetonia Creek, a tributary to the Little Washita River.

During this 1978-1992 period, the water quality research shifted from the LWREW to include other watersheds across Oklahoma and the Southern Great Plains. Hydrologic research on the LWREW changed from predominantly observational and experimental research to distributed hydrologic model applications to quantify runoff volume, peak flow rates, sediment yield, and water quality parameters. The hydrologic modeling approach was found to be well suited for assessing long term effects of major land treatment shifts and flood retarding structures on the hydrology and sediment yield of the LWREW. In the early 1990s, model applications included investigation of the impact of global climate change on watershed hydrology.

Research was also conducted on the impact of observed climatic variations on watershed runoff hydrology and on the effectiveness of flood retarding structures to reduce peak flow rates. This research demonstrated the important role climatic variations and trends play in watershed hydrology and soil and water conservation.

Little Washita River Experimental Watershed, 1992-2004
The 1992-2004 period was a time of great change and new opportunities for the LWREW. In the early 1990s, the ARS-Global Change, Water Resources and Agriculture (ARS-GCWRA) program gave impetus for re-activation and instrumentation upgrades on the LWREW. This new research emphasized global change issues and addressed the exchange of water and energy in managed ecosystems, and the effects of land cover and climate on land-surface hydrology. This led to 8 multi-agency, multi-disciplinary, intensive field campaigns which took place on the LWREW that are described in a separate section.

The re-instrumentation of the LWREW began in 1992 and consisted of a network of 42 climate stations known as the "Micronet" (Figure 4). Most stations were placed on the 3 by 3 mile grid of the 1978-1985 rain gauge network. These climate stations measured rainfall, air temperature, relative humidity, incoming solar radiation, and soil temperature at four depths. In addition, three climate stations with complemetary instrumentation, called "Mesonet stations" and operated by the Oklahoma Climatological Survey, were located on or near the LWREW.

Starting in 1995, selected Micronet sites were complemented with measurements of soil water content, soil temperature, and soil heat flux (SHAWM sites in Figure 4). The U.S. Department of Energy (DoE), the National Oceanic and Atmospheric Administration (NOAA) and the Natural Resources Conservation Service (NRCS) also installed heat and moisture flux measurement sites.

In 1992, three stream gauging stations were installed by the USGS along the main stem of the Little Washita River. Three more stations were added in 1995 on tributaries to the Little Washita River and a fourth one was added in 1996. In 1993, 1995, and 1996, three reservoir-level gauging stations were added to the network to monitor the pool elevation of SCS flood retarding structures.

Data from this dense observational network have been used by the National Weather Service and researchers from the University of Oklahoma to improve and verify NEXRAD radar algorithms for rainfall estimation, and by NOAA's River Forecast Center in Tulsa, Oklahoma, to refine flood prediction models. These applications led to spatially-distributed precipitation products that are used in large-scale hydrologic modeling and flood forecasting. The LWREW data and observation capabilities also attracted numerous, inter-agency watershed-scale hydrologic experiments (see separate section) that enabled development of new remote sensing systems and associated algorithms for surface soil moisture estimation over large land areas. Soil water measurements in the LWREW have been: (1) used to validate land-vegetation-atmosphere moisture-transfer models; (2) incorporated into a simplified soil hydrology model to simulate soil water content in the soil profile; and, (3) used to verify soil moisture estimates from radar sensors aboard a NASA satellite.

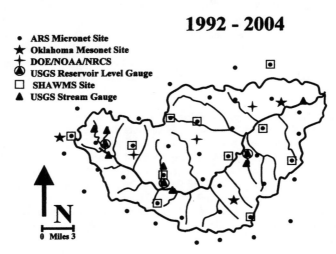

Figure 4. The Little Washita River Experimental Watershed 1992-2004.

Little Washita River Experimental Watershed, 2004-2005

In 2003, the Conservation Effects Assessment Project (CEAP) was initiated. CEAP is a Natural Resources Conservation Service (NRCS) led study designed to quantify environmental benefits of conservation practices implemented on agricultural lands (CEAP, 2005). The low percentage of cropland on the LWREW (about 18%) provided limited opportunities to address immediate CEAP objectives. This drawback was remedied by redirecting the research focus to the nearby Fort Cobb Reservoir Experimental Watershed (FCREW) (see next section).

In support of the CEAP effort, 15 climate stations were relocated from the LWREW to the FCREW. In addition, five climate stations, two reservoir-pool elevation gauges and three stream gauging stations were discontinued on the LWREW, leaving 20 climate stations, one reservoir-pool elevation gauge, and 4 stream gauging stations. The 20 climate stations were retrofitted with new soil moisture sensors that replaced the old SHAWM instruments. These remaining monitoring capabilities on the LWREW are sufficient to support ongoing water resources research involving hydrologic modeling, as well as anticipated water quality research needs associated with the upcoming extension of CEAP efforts to rangeland watersheds. In 2003 and through 2005, hydrologic model application studies were conducted on the LWREW to test model performance, quantify impacts of seasonal and multi-year climate variations, and explore opportunities presented by seasonal climate forecasts for water resources management.

Fort Cobb Reservoir Experimental Watershed, 2004-2005
The FCREW has been identified by the Oklahoma Water Resources Board, the Oklahoma Conservation Commission, and the Oklahoma NRCS as a focal point to apply conservation practices to land in the watershed in order to improve water quality in the reservoir. The Fort Cobb Reservoir and watershed have been the focus of monitoring and assessments for a number of years, providing a baseline of data against which future environmental conditions can be assessed. The objective of ARS' Fort Cobb CEAP study is to assess the effects and benefits of selected conservation practices as they relate to reduction of inputs of suspended sediments to surface water, and the reduction of phosphorus and nitrogen in surface and ground water.

Precipitation and other weather variables on the FCREW have been monitored by three institutions: since 1994, three Mesonet climate stations have been operated by the Oklahoma Climatological Service; four Cooperative Weather Stations with over 20 years of data have been operated by the National Weather Service; and, in 2005, 15 Micronet climate stations were relocated from the LWREW to the FCREW and are operated by ARS (Figure 5). Stream flow on Cobb Creek has been monitored by the USGS since 1939. In 1968, 1969, and 1970, three USGS stream gauges were installed on tributaries to Cobb Creek. A fourth nested stream gage was added in 2005 (Figure 5). At three of the tributary gauging stations, discharge and event-based sediment and water quality have been monitored since 2004.

Initial CEAP-related research efforts focused on the development of spatially-distributed land use and physiographic data bases for hydrologic model applications, and on model calibration and validation for FCREW conditions. First model applications quantified the decrease in sediment yield reduction due to upstream conservation measures with downstream distance along the channel. A separate evaluation of observational data uncovered a strong sensitivity of runoff and particularly sediment yield to multi-year precipitation variations. Based on these research findings, recommendations with regard to model calibration and assessment of conservation measures under variable climate conditions were made.

Figure 5: The Fort Cobb Reservoir Experimental Watershed (2004-2005) climate stations (Micronet).

Major Hydrologic Field Campaigns on the LWREW between 1992 and 2003

STORM-FEST
STORM-FEST was a United States Weather Service research project concerning winter storms and frontal evolution in southcentral Oklahoma and a systems test of new radar technologies. The LWREW was chosen as the hydrology experiment site for STORM-FEST because of the historical climate/weather/hydrology database on the watershed and its proximity to three nearby WSR-88D radars (Weather Service Radar-1088 Doppler or NEXRAD).

WASHITA '92
Washita '92 was a cooperative experiment that emphasized the collection and analysis of remotely sensed data for hydrologic investigations. This experiment involved USDA-ARS, NASA, and several other government agencies and universities. Data collection during Washita '92 emphasized determination of soil moisture and evaporative fluxes, using both conventional ground sampling and remote sensing techniques. A second objective was the testing and evaluation of new microwave remote sensing devices for soil moisture measurement.

Remotely sensed data were collected in a multi-altitude fashion. A NASA C-130 aircraft provided a low to medium altitude platform for remote sensing instruments. A DC-8 aircraft provided a high altitude platform for NASA's Synthetic Aperture Radar (SAR). Ground level remotely sensed data were acquired by a radar mounted on the NASA Goddard Space Flight Center truck and hand-held instrumentation. Soil samples were collected daily at a large number of predetermined sites during the experiment to provide estimates of soil water content over the study area. This experiment produced one of the most widely studied soil moisture remote sensing data sets that resulted in broad impact in hydrologic sciences.

WASHITA '94

Like Washita '92, this short term intensive field experiment involved many Federal and State research agencies and institutions and was conducted in three phases. The first phase concentrated on the remote sensing of soil moisture in the LWREW. The NASA C-130, DC-8 and Space Shuttle Endeavour collected data over selected portions of the watershed using instrumentation similar to that deployed during the Washita '92 campaign. The second phase, conducted 4 months later, was similar to the first phase except that it did not include NASA flights. Remotely sensed data were obtained for water quality studies and were acquired during overflights by the ARS Aero Commander aircraft using video cameras. Ground crews collected water samples and remotely sensed data from the impoundments near the time of the aircraft overflights. The third phase, conducted in November of 1994, utilized microwave instrumentation aboard the Space Shuttle Endeavour for soil moisture determinations. Analysis of the Space Shuttle Imaging Radar data resulted in the first robust radar soil moisture algorithm.

WASHITA '95

During the summer of 1995 the ARS aircraft overflew the LWREW to study water quality of small to medium-sized impoundments. Optical sensors including video cameras and radiometers were equipped with special filters to detect the presence of algae and suspended sediments. Water samples were collected from the impoundments and analyzed for suspended sediments and the presence and amount of algae. Results from this study demonstrated the feasibility of detecting and quantifying chlorophyll and suspended sediment using low-cost videography flown on board low-altitude aircraft.

WASHITA '96

A joint ARS/NOAA soil moisture study was conducted in the summer of 1996. This effort included overflights of the LWREW by NOAA aircraft equipped with a gamma radiation sensor to detect the presence and amount of soil moisture along transects in the LWREW. An intensive field measurement campaign was conducted to verify results from the gamma radiation sensor.

SGP '97

The main objective of the Southern Great Plains 1997 (SGP '97) Hydrology Experiment was to establish whether or not retrieval algorithms for surface soil moisture developed from high spatial resolution truck- and aircraft-based sensors could be extended to coarser resolutions expected from future satellite platforms. The core of the experiment involved deployment of microwave radiometers for daily mapping of surface soil moisture over an area greater than 4,000 square miles and a period of approximately one month. Numerous research teams from State and Federal research agencies, universities, national and international, and foreign research laboratories converged on Oklahoma to address additional soil moisture and temperature objectives.

SGP '99

The SGP '97 Hydrology Experiment successfully demonstrated the ability to map and monitor soil moisture using low frequency microwave radiometers. Soil moisture retrieval algorithms developed using high spatial resolution data were proven to be extendible to coarser spatial resolutions. The major goals for the Southern Great Plains 1999 Experiment (SGP '99) were: (1) to understand how to effectively interpret and use satellite microwave data; (2) explore new approaches that may enhance the ability to measure soil moisture from space, and (3) to provide an understanding of surface temperature variability in the study region for interpreting thermal sensor observations.

SMEX03

The Soil Moisture Experiment in the 2003 (SMEX03) project was conducted at various sites in Oklahoma, Georgia, Alabama and Brazil. The purpose of this experiment was to validate the soil moisture products derived from the NASA Aqua and Japanese Advanced Microwave Scanning Radiometers, which provided daily soil moisture products for a variety of landcover conditions. Because of the large footprint (about 15 square miles), well-instrumented watersheds like the LWREW were needed to augment the traditional groundtruthing efforts typical of earlier hydrology experiments.

Concluding Comments

The three USDA-ARS experimental watershed on the Washita River have served for over 40 years as an important outdoor laboratory for assessing environmental impacts of flood retarding structures, soil and water conservation practices, and land management in the Southern Great Plains. Four major research periods and themes are recognized. First, the 1961-1978 period dealt with the assessment of the impact of flood retarding structures on watershed hydrology and sediment yield. Second, the 1978-1985 period focused on control of non-point source pollution on water quality using best management practices in large watersheds. Third, the 1992-2004 period addressed potential hydrologic impacts of global climate change and development of remote sensing technologies for large-scale soil moisture measurements. And, fourth, the 2004-present period emphasizes the development and evaluation of procedures for assessing the environmental and societal benefits associated with federally funded conservation practices (CEAP). Much of the watershed data supported multi-agency, multi-disciplinary research involving flood reduction, soil and water conservation, agricultural water quality, sedimentation, erosion, energy fluxes, soil moisture, and remote sensing. Investigations of watershed process on the LWREW and FCREW also recognized early-on the need to link watershed research across a range of scales, and to target non-point source controls and best management practices to critical source areas. Long term research is essential to properly evaluate land management and climate on water quantity and quality impacts. Expectations are that the Little Washita River Watershed, together with the more recent Fort Cobb Watershed, will continue to provide valuable land use information for protecting and utilizing soil and water resources in the Southern Great Plains.

Acknowledgments

This brief history of the Little Washita River and Fort Cobb Reservoir Experimental Watersheds would not have been possible without the help from those who have built, managed and worked with these research watersheds. The authors recognize their outstanding and sustained contributions to experimental watershed sciences, and, in particular, their contributions to this historical review. Special thanks are extended to G. A. Coleman, T. J. Jackson, J. W. Naney, F. R. Schiebe, R. R. Schoof, A. N. Sharpley, S. J. Smith, and R. D. Williams for their contributions to this paper. The authors are also grateful to J. A. Daniel and M. W. Van Liew for contributing valuable background information pertaining to groundwater and hydrology aspects of the watersheds.

References

Allen, P. B. and J. W. Naney. 1991. Hydrology of the Little Washita River Watershed, Oklahoma. U. S. Dept. of Agriculture, Agricultural Research Service, Publication No. ARS-90, 74 pp.

Browning, G. M., G. E. Ryerson, C. H. Wadleigh, and D. M. Whitt. 1959. A report of findings by the Working Group appointed by the Secretary of Agriculture, USDA, January 1959, presented as Document 59 by Mr. Hayden to the 86th Congress.

CEAP. 2005. Conservation Effects Assessment Project – The ARS Watershed Assessment Study, Project Plan 2005. Available at: http://www.nrcs.gov/technical/NRI/ceap/index.html#.

Staff, Water Quality and Watershed Research Laboratory. 1983. Hydrology, erosion, and water quality studies in the Southern Great Plains Research Watershed, Southwestern Oklahoma. 1961-1978. U. S. Dept. of Agriculture, Agricultural Research Service, Publication No. ARM-S-29, 175 pp.

History of the USDA-ARS Walnut Gulch Experimental Watershed

M. H. Nichols[1] and K. Renard[2]

[1]USDA ARS Southwest Watershed Research Center, 2000 E Allen Rd, Tucson, AZ 85719; PH (520)670-6381; FAX (520)670-5550; email: mnichols@tucson.ars.ag.gov
[2]USDA ARS Southwest Watershed Research Center, 2000 E Allen Rd, Tucson, AZ 85719; PH (520)670-6381; FAX (520)670-5550; email: krenard@tucson.ars.ag.gov

Abstract

The United States Department of Agriculture Agricultural Research Service Walnut Gulch Experimental Watershed (WGEW) has served as an outdoor laboratory for the study of semi-arid hydrology, erosion and sedimentation since the 1950s. This paper presents an overview of the core WGEW instrumentation network, its development, and evolution. The need for specialized instruments to measure high velocity, short duration, sediment laden flow that characterize semi-arid regions and the development of the Walnut Gulch Supercritical Depth Runoff Measuring Flume are described. Recent upgrades to the instrumentation network include conversion from analog to digital recording methods, automated radio-based data transfer from sensors to a centralized computer network, and access to data via the Internet. The WGEW is a critical resource for enhancing our scientific understanding of semi-arid hydrology and ecosystems and the impact of their interactions on water supply and quality.

Introduction

The southwestern U.S. is characterized by its semi-arid climate. Until the mid 1900s data and information to quantify the relationships among rainfall, runoff, land use, and land management in semi-arid rangelands were lacking. In 1953, research on the United States Department of Agriculture Agricultural Research Service's Walnut Gulch Experimental Watershed (WGEW) was initiated with a principal objective to determine the impact of upland conservation practices on water yields and sediment loads (Renard et al., in review).

The 150 km² WGEW surrounds the town of Tombstone in southeastern Arizona (Figure 1). The WGEW site was chosen based on an exhaustive regional search to select an area suitable for long-term hydrologic, range management, and erosion research. Historically the area has supported grazing and mining, and recently recreation has become a dominant land use. The WGEW is located in the Basin and Range Province (Austin, 1981) in the semi-arid transition zone between the Sonoran and Chihuahuan Deserts. The watershed drains westward, and the main Walnut Gulch channel joins the San Pedro River near Fairbanks, Arizona. The San Pedro River originates in Sonora, Mexico and flows north into and through Arizona.

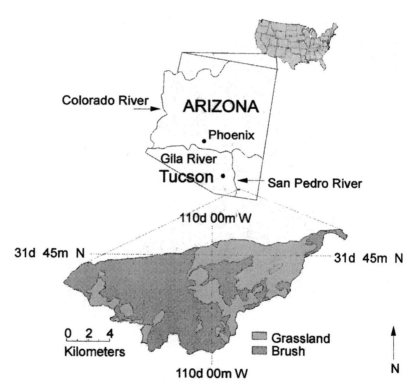

Figure 1. USDA-ARS Walnut Gulch Experimental Watershed location map.

Instrumentation

The first instruments installed on the watershed were raingages and flumes
that would supply data to support quantifying flood hydrology. The raingage network
was initiated with 11 raingages and has expanded to the current network consisting of
almost 100 raingages (Goodrich et al., in review). The high number of raingages is
necessary to accurately quantify rainfall generated by highly localized thunderstorm
cells that are common during July, August, and September.

During 1953, 5 large, concrete, trapezoidal critical depth runoff measuring
flumes were constructed. The hydraulic sections of these flumes were in place by July
of 1954 with design capacities ranging from 42.5 to 226.5 cms. Significant rainfall on
the watershed during the first summer of operation yielded runoff that resulted in the
failure of all of the flumes. The flumes failed because they were 1) undersized, 2)
hydraulically under designed, and 3) structurally incapable of withstanding the forces
of the flood flows.

These failures led to the development of the Walnut Gulch Supercritical
Runoff Flume (Gwinn, 1964; Gwinn, 1970; Smith et al., 1982). The flume was
designed in cooperation with the ARS Hydraulics Laboratory in Stillwater, OK where
scale models of the channel and flumes were tested. Flow measurement in ephemeral
channels such as Walnut Gulch is complicated by heavy sediment loads. The new
flume was designed to discharge all of the sediment that enters and is thus self-
cleaning. A second, smaller (2.8 cms), critical depth flume (called the Smith flume or
Santa Rita flume) was designed to measure runoff from small watersheds. Runoff
measurements to characterize volume, peak discharge rate, and flow duration have
been collected since the early 1960s. The resulting runoff database comprises the
longest temporal sequence of runoff within and through a semi-arid watershed in the
world (Stone et al., in review).

Flows on the WGEW are infrequent, but reach high velocities and carry heavy
sediment loads. V-notch weirs developed and used in more humid environments did
not work well on the WGEW because sediment deposition resulted in the loss of
hydraulic control required for accurate runoff measurement and sediment sampling.
Sediment measurement required the development of new instrumentation to
accommodate sediment loads made up of a broad range of particle sizes. The
traversing slot sediment sampler was designed in response to limitations of
alternative sampling methods (Nichols et al., in review, Renard et al., 1986).

Expanded instrument network

Since the 1950s, research objectives have expanded to include new
technologies, such as remote sensing and simulation modeling to address complex
questions that integrate broader concerns such as climate impact, land use and
demographics, and resource management. Currently, a wide range of sensors are
deployed to collect air temperature, relative humidity, wind speed, wind direction,

solar radiation, net radiation, photosynthetically active radiation, and barometric pressure. In addition, carbon flux, soil moisture, soil temperature, soil-heat flux and soil-surface temperature measurements are collected (Keefer et al., in review).

Operation, maintenance, and data collection at WGEW are costly and labor intensive. The mechanical clocks and analog charts used in the original rainfall and runoff instruments were becoming increasingly obsolete by the 1990s. In 1996, an effort to fully re-instrument the WGEW with electronic sensors and digital data-loggers was initiated. The instrument upgrade was paralleled by the installation of radio telemetry equipment to remotely transmit recorded data to a central computer. The instrumentation upgrade has greatly reduced operational overhead by reducing labor, maintenance, and data. Currently, there are 125 core instrument installations on the WGEW that provide data on a daily basis (Figure 2).

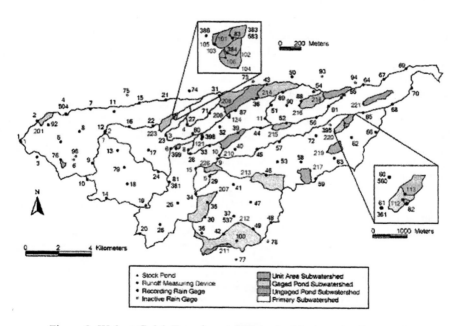

Figure 2. Walnut Gulch Experimental Watershed instrumentation map.

Summary

Long-term research programs are critical to the conservation of semi-arid lands. Research at the WGEW is ongoing in cooperation with the USDA Natural Resource Conservation Service, local Natural Resource Conservation Districts, local ranchers, universities, and international scientists interested in understanding semi-arid watersheds. Research on the WGEW has resulted in the development of new instrumentation which continues to evolve in response to changing technologies. As a result the WGEW is the most highly instrumented semi-arid experimental watershed in the world. The measurement infrastructure and the resulting spatially and temporally distributed natural resources database are of national and international importance for addressing soil and water concerns through semi-arid regions.

References

Austin, M.E. (1981). "Land Resource Regions and Major Land Resource Areas of the United States." *USDA, Agricultural Handbook No. 296.*

Goodrich, D.C., T.O. Keefer, C.L. Unkrich, M.H. Nichols, H.B. Osborn, J.J. Stone, and J.R. Smith. (2007). "Long-term precipitation database, WGEW, Arizona, USA." *Water Resources Research,* in review

Gwinn, W.R. (1964). "Walnut Gulch Supercritical Flumes." *Proc. Am. Soc. Agric. Engrs,* 10(3), 197-199.

Gwinn, W.R. (1970). "Calibration of Walnut Gulch Supercritical Flumes." *Proc. Am. Soc. Civil Engrs,* 98(HY8), 1681-1689.

Keefer T. O., M.S. Moran and G.B. Paige. (2007). "Long-term meteorological and soil-dynamics database, Walnut Gulch Experimental Watershed, Arizona, USA." *Water Resources Research,* in review.

Nichols, M., J.J. Stone and M.A. Nearing. (2007). "Long-term sediment database, WGEW, Arizona, USA." *Water Resources Research,* in review

Renard, K., M. Nichols, D. Woolhiser and H. Osborn. (2007). "The history of ARS watershed research and modeling in Arizona and New Mexico." *Water Resources Research,* in review.

Renard, K.G., Simanton, J.R., and Fancher, C.E. (1986). "Small watershed automatic water quality sampler." *Proceedings of the 4th Federal Interagency Sedimentation Conference, Las Vegas, NV,* vol.1, pp51-58.

Smith, R.E., D.L. Chery, Jr., K.G. Renard and W.R. Gwinn. (1982). "Supercritical flow flumes for measuring sediment-laden flow." *USDA-ARS, Technical Bulletin Number 1655*, Washington D.C., 70 pp.

Stone, J.J., M. Nichols, D.C. Goodrich and J. Buono. (2007). "Long-term runoff database, WGEW, Arizona, USA." *Water Resources Research,* in review

Goodwin Creek Experimental Watershed:
A Historical Perspective

R. L. Bingner[1], R A. Kuhnle[2], C. V. Alonso[3]

[1]National Sedimentation Laboratory, United States Department of Agriculture – Agricultural Research Service, P.O. Box 1157, Oxford, MS 38655; PH (662) 232-2966; FAX (662) 281-5706; email: RBingner@ars.usda.gov
[2] National Sedimentation Laboratory, United States Department of Agriculture – Agricultural Research Service, P.O. Box 1157, Oxford, MS 38655; PH (662) 232-2971; FAX (662) 281-5706; email: RKuhnle@ars.usda.gov
[3] National Sedimentation Laboratory, United States Department of Agriculture – Agricultural Research Service, P.O. Box 1157, Oxford, MS 38655; PH (662) 232-2969; FAX (662) 281-5706; email: CAlonso@ars.usda.gov

Abstract

The Goodwin Creek Experimental Watershed (GCEW) was established in north central Mississippi by U.S. Congressional action and the U.S. Department of Agriculture (USDA) National Sedimentation Laboratory (NSL) has operated the watershed since October, 1981. Since then, the watershed has provided a platform for research ranging from watershed hydrology, stream fluvial studies for sediment transport relationship development and channel evolution, stream restoration, landuse effects on sediment load, sediment source identification, climatic changes, and watershed and channel model development. This paper will discuss the historical perspective and national impact of this climatological, hydrological, fluvial, and sedimentation research collected at the watershed.

Introduction

U.S. Public Law 93-251, Section 32, established the Goodwin Creek Experimental Watershed (GCEW) in north central Mississippi (Figure 1) as a part of a "Streambank Erosion Control Evaluation and Demonstration Project" (DEC). This area of Mississippi has been characterized as having excessive upland erosion, steep degrading channels, loss of land due to channel bank caving, and downstream deposition problems, which are characteristics of many watersheds throughout the U.S. Construction funds for much of the watershed infrastructure were originally provided by the U.S. Army Corps of Engineers (USACE), Vicksburg District. The U.S. Department of Agriculture (USDA) National Sedimentation Laboratory (NSL)

113

has continuously monitored the watershed since October, 1981.

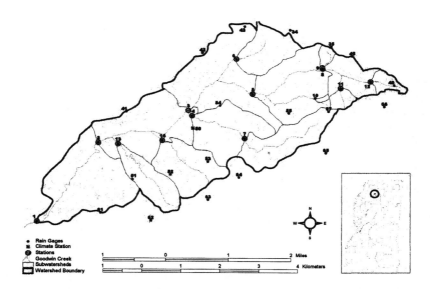

Figure 1. Goodwin Creek Experimental Watershed gaging system.

The data collected within GCEW has included: raingages; air temperature; solar radiation; soil moisture and soil temperature; runoff; sediment; nutrients; channel characterization through repeated cross-section surveys, bendway studies, vegetated stream bank effects, particle size distribution surveys of the bed, and stone weirs for wildlife restoration; and spatial and temporal changes in landuse throughout the watershed. Data has been digitized for utilization in geographic information systems (GIS) that describe the spatial variability of landuse, soils, topography and other features such as the location of gages, roads, channels and other significant landscape features. This database has served as a platform for many research efforts in the understanding of watershed processes, and for the development and validation of models that incorporate the description of these processes (Blackmarr, 1995). The database is available for download, along with the documentation, from the Website: **http://ars.usda.gov/Business/docs.htm?docid=5040.**

In addition to the DEC project, GCEW has been involved with several other national coordinated research efforts with their objectives to describe various watershed hydrologic components or effects of conservation practices. The USDA Conservation Effects Assessment Project (CEAP) selected GCEW in 2003 as one of the original twelve, and currently fourteen, USDA Agricultural Research Service (ARS) benchmark watersheds that will be used to evaluate the effects of conservation practices on water quality at a watershed scale. Ongoing research on weather radar applications to rainfall measurement has been aimed at showing how to combine weather radar (NEXRAD), raingauge, and raindrop spectrometer information to provide rainfall rate and rainfall kinetic energy concurrently, while covering areas of a large scale at high spatial and temporal resolution (Steiner et al., 1999). One of the most important potential benefits of using weather radar can be to augment sparse gauge networks, such as in developing countries, to provide comprehensive coverage of major storms passing over areas subject to flooding or erosion.

Since December 1994, the GCEW has been part of the national network of Surface Radiation Budget (SURFRAD) stations operated by NOAA for long-term measurement of the earth surface radiation budget (DeLuisi et al., 1996). The data records at all SURFRAD stations are available on the SURFRAD Website: **http://www.srrb.noaa.gov/surfrad/index.html**. The GCEW also participates in the NRCS Soil Climate Analysis Network (SCAN) that monitors soil moisture and soil temperature throughout the United States. Two SCAN stations were installed in January of 1999 on pasture and timber areas within GCEW. These stations provide continuous soil water measurements at depths of 5, 10, 20, 51, and 102 cm. Data from the SCAN and SURFRAD networks play a vital role in validating satellite scanners and developing better understanding of land-atmosphere interactions and the water balance itself (Jackson, 1997). The data can be downloaded from the SCAN Website at: **http://www.wcc.nrcs.usda.gov/scan/index.html.**

The ARS role in DEC, CEAP, NEXRAD, SCAN, and SURFRAD data includes developing improved procedures for assessing the impact of conservation measures on watershed water quality and in determining the severity and frequency of occurrence of extreme weather events and long-term trends in weather patterns, better soil-water-crop and hydrological relationships, drought mitigation, agricultural production, and model process scaling from point to global.

GCEW Data System

In-Stream Flumes. The watershed area was originally partitioned into 14 nested subbasins so it would not have more than 1/3 of its contributing area ungaged (Figure 1). Measuring flumes were constructed at each subbasin outlet, and the drainage areas above these stream gaging sites range from 0.053 to 21.3 km^2. The flumes served the dual purposes of controlling degradation of the channel bed, and measuring runoff and sediment loads.

Raingages. Continuous precipitation data are collected at 30 recording raingages located at sites both inside and outside the watershed area (Figure 1). Weather and climatological data are collected from a hydrometeorological station located near the center of the watershed.

Telemetry System. Monitoring has been continuously improved to incorporate a data acquisition, real-time telemetry system from all field sites that leads to all data incorporated into a relational data base for archiving.

Landscape Characterization. The watershed's topography is documented by 15- and 7.5-minute US Geological Survey Quadrangle maps, USGS Digital Elevation Models, a 1:5000 scale contour-interval relief map, and 1:500 scale detailed channel surveys prepared by the Corps of Engineers. The USDA-Natural Resources Conservation Service has classified and mapped the soil associations in the watershed. Cross-sectional surveys of Goodwin Creek have been conducted since 1977 to document changes in the channel geometry and to verify the sediment delivery budget of the channel system. Since 1980, periodic ground and aerial photography surveys were conducted to characterize crop and cover conditions in the watershed and how these conditions change over time in each field. Currently, less than 13 percent of the total watershed area is under cultivation, with the rest defined as 10% idle, 50% in pasture, and 27% forested. Twenty-five years of continuous, analyzed hydrologic data are now available for Goodwin Creek.

Discussion

The research and data collection effort within GCEW has provided a unique perspective on understanding watershed processes, the impact conservation measures can have on sedimentation, and the development of effective watershed management plans. Long-term studies at GCEW have provided an avenue to study the entire watershed system from fields to channels and from single storm events to climatic cycles that may be occurring. Studies have shown that the instream flumes have helped stabilize the channel, while streambank restoration projects have helped stabilize the banks and reduce erosion (Bowie, 1995).

Sediment data collected on the GCEW indicate that mean annual concentration of sediment in the main channel decreased by 60% from 1982 to 1990 and that this change was primarily caused by the change in land use that occurred over this period (Kuhnle et al., 1996). The mean annual sediment concentrations have been essentially stable since about 1990 (Kuhnle et al., 2006). On average, it has been established from measured sediment data, results from watershed models, and sediment tracking using naturally occurring radionuclides, that approximately 80% of the sediment transported by the main channel of the GCEW originated from channel sources (Kuhnle et al., 1996; Wilson and Kuhnle, 2006).

Significant watershed technology has been developed through the utilization of data and research from GCEW, such as the Annualized Agricultural Non-Point Source (AnnAGNPS) pollution model (Bingner and Theurer, 2001) and the Conservational Channel Evolution and Pollutant Transport System (CONCEPTS) (Langendoen, 2000). The application of models provides an avenue to assess the effect of conservation practices used in watershed management plans. Both AnnAGNPS and CONCEPTS have been designated for use in ARS benchmark and NRCS special emphasis watersheds in the CEAP project.

After 25 years of research on GCEW, the combination of an intensively instrumented watershed with targeted studies has provided an important outdoor laboratory to understand how watershed processes are integrated together. When the system is out-of-balance from erosive agricultural practices or from a disturbance in channels as a result of construction or straightening, the downstream impacts may not be known for many years or decades later. The effect of agricultural and channel conservation measures are not easily quantified, with a need for many more years of research to fully realize the additional effect of climate change on the watershed system.

References

Bingner, R. L. and Theurer, F. D. (2001). AnnAGNPS: estimating sediment yield by particle size for sheet & rill erosion. In Proceedings of the Sedimentation: Monitoring, Modeling, and Managing, 7th Federal Interagency Sedimentation Conference, Reno, NV, 25-29 March 2001. p. I-1 - I-7.

Blackmarr, W.A. (Ed.). (1995). Documentation of hydrologic, geomorphic, and sediment transport measurements on the Goodwin Creek Experimental Watershed, Northern Mississippi, for the period 1982-1993, Preliminary Release. Research Report No. 3, USDA-ARS National Sedimentation Laboratory, Oxford, Mississippi, October 1995, 216 pp.

Bowie, A. J. (1995). Use of vegetation to stabilize eroding streambanks. U.S. Department of Agriculture, Conservation Research Report 43, 24 pp.

DeLuisi, J. J., J. A. Augustine, E. C. Weatherhead, B. B. Hicks, D. Matt, and C. V. Alonso. (1996). The GCIP Surface Radiation Budget Network (SURFRAD). Proceedings, Second International Scientific Conference on the Global Energy and Water Cycle, 407-408, Washington, DC, June 17-21, 1996.

Jackson, T.J. (1997). Soil moisture estimation using special satellite microwave/imager satellite data over a grassland region, Water Resources Research, 33(6):1475-1484.

Kuhnle, R. A., Bingner, R. L., Foster, G. R. and Grissinger, E. H. (1996). Effect of land use changes on sediment transport in Goodwin Creek. Water Resources Research 32(10): 3189-3196.

Kuhnle, R. A., Bingner, R. L., Alonso, C. V., Wilson, C. G., 2006. Goodwin Creek Experimental Watershed - Effect of Conservation Practices on Sediment Load. ASABE paper no. 062069.

Langendoen, E.J. (2000). CONCEPTS—Conservational Channel Evolution and Pollutant Transport System. Research Report No. 16, USDA-ARS National Sedimentation Laboratory, Oxford, MS. 160 pp.

Steiner, M., J.A. Smith, S.J. Burgess, C.V. Alonso, and R.W. Darden. (1999). Effect of bias adjustment and rain gauge data quality control on spatial radar rainfall estimation, Water Resources Research, 35(8):2487-2503.

Wilson, C. G., Kuhnle, R. A., 2006. Determining relative contributions of eroded landscape sediment and bank sediment to the suspended load of Goodwin Creek using 7Be and 210Pbxs. National Sedimentation Laboratory, Technical Report No. 53.

History of Hydrologic Research and Data Collection at the
Grassland, Soil and Water Research Laboratory, Riesel, Texas[1]

R.D. Harmel[2], and C.W. Richardson[3]

[1] Mention of trade names or commercial products is solely for the purpose of
providing specific information and does not imply recommendation or endorsement
by the USDA.
[2] Agricultural Engineer, USDA-ARS, 808 E. Blackland Road, Temple, TX, 76502;
PH (254) 770-6521; email: dharmel@spa.ars.usda.gov.
[3] Former Laboratory Director, USDA-ARS, 808 E. Blackland Road, Temple, TX,
76502; PH (254) 770-6517; email: crichardson@spa.ars.usda.gov.

ABSTRACT
The history of hydrologic research at the Riesel, TX, site dates to the mid 1930's
when the United States Department of Agriculture Soil Conservation Service
(USDA-SCS) realized a need to analyze and understand hydrologic processes of
agricultural fields and watersheds and their impact on soil erosion, flood events,
water resources, and the agricultural economy. As a result, the Blacklands
Experimental Watershed was established in 1937 and remains in operation today as
part of the USDA Agricultural Research Service (USDA-ARS) Grassland, Soil and
Water Research Laboratory. The Riesel watershed (as well as experimental
watersheds located near Hastings, NE, and Coshocton, OH) were established within
the research program of USDA-SCS to collect hydrologic data and to evaluate the
hydrologic and soil loss response as influenced by various agricultural land
management practices. A major contribution of the original three watersheds is the
quantification of soil loss reduction under conservation management, which has led to
drastic reduction in soil loss from cultivated agriculture since the early 20th century.
More recently, research at Riesel established fundamental understanding of the
agronomic and environmental effects of tillage, fertilizer, and chemical alternatives.
Riesel watershed data and analyses also produced the scientific basis for several
watershed models that are now used worldwide to manage water quality. An
enhanced understanding of Texas Blackland Prairie hydrology has also resulted
because of the continuous, long-term hydrologic data record and associated analyses.
Because of their historical and future value, USDA-ARS has a unique responsibility
to maintain such long-term watershed research sites to help meet future issues related
to water availability, environmental quality, and food and fiber production.

INTRODUCTION
In the mid 1930's, the United States Department of Agriculture Soil Conservation Service (USDA-SCS), now the Natural Resources Conservation Service (USDA-NRCS), realized a need to analyze and understand hydrologic processes on agricultural fields and watersheds and their impact on soil erosion, flood events, water resources, and the agricultural economy. At that time, the Hydrologic Division was established and organized as part of the SCS research program, and provisions were made for establishing a number of experimental watersheds (USDA-ARS, 1942). As a result, SCS established experimental watersheds in Coshocton, OH, Riesel, TX, and Hastings, NE. These original watersheds were designed to collect hydrologic data (precipitation, percolation, evaporation, runoff) and to evaluate the hydrologic and soil loss response as influenced by various agricultural land management practices. The Blacklands Experimental Watershed near Riesel, TX, (Figure 1) has been in continuous operation since 1937, although the operation was scaled back during World War II. The original three experimental watersheds have been managed and operated by the USDA Agricultural Research Service (USDA-ARS) since its formation in 1954. The Riesel watershed has been known as the Grassland, Soil and Water Research Laboratory since 1977 when the laboratory headquarters in Temple, TX, were renamed as such. In addition to the three original watersheds, seven USDA-ARS experimental watersheds were established from 1953-1980 as the result of legislation such as the Senate Document 59 (86th Congress, 1959) and Public Law 93-251.

Site Description
The Grassland, Soil and Water Research Laboratory experimental watershed, which currently contains 340 ha of federally owned and operated land, was established on the 2372 ha Brushy Creek watershed near Riesel, TX, because of its central location in the 4.45 million ha Texas Blackland Prairie (Figure 1). Present day agricultural land use in this productive region consists of cattle production on pasture and rangeland, and corn, wheat, grain sorghum, and oat production under a wide range of tillage and management operations. The area also contains major portions of Dallas-Fort Worth and Austin, which are major metropolitan areas. Houston Black clay soils (fine, smectitic, thermic, udic Haplustert), recognized throughout the world as the classic Vertisol, dominate the watershed site. These highly expansive clays, which shrink and swell considerably with changes in moisture content, have a typical particle size distribution of 17% sand, 28% silt, and 55% clay. This soil series consists of very deep, moderately well-drained soils formed from weakly consolidated calcareous clays and marls and generally occurs on 1-3% slopes in upland areas. This soil is very slowly permeable when wet (approximate saturated hydraulic conductivity of 1.5 mm/hr); however, preferential flow associated with soil cracks contributes to high infiltration rates when the soil is dry (Arnold et al., 2005; Allen et al., 2005; Harmel et al., 2006).

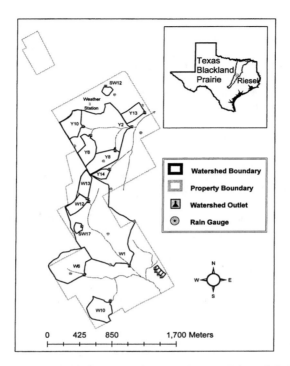

Figure 1. The USDA-ARS Grassland, Soil and Water Research Laboratory
experimental watershed, near Riesel, TX.

HYDROLOGIC INSTRUMENTATION AND DATA COLLECTION
The current hydrologic instrumentation network at Riesel includes 17 runoff and
water quality stations, 15 rain gauges, a weather station, a lateral flow station, and
seven shallow groundwater wells. Data from these sites are updated periodically and
placed on the internet with data from additional historically-active sites. In addition
to hydrological and meteorological data, land management data, such as tillage
operations, fertilizer application, and crop yields are available for many fields and
watersheds. Historical data for active and inactive sites are freely available at
www.ars.usda.gov/spa/hydro-data.

Runoff and Water Quality Stations
Discharge data collection began in 1938 and has been conducted at various times
from 40 watersheds ranging from 0.1-2372 ha. A hydraulic (flow control) structure is
used at the outlet of each of these stations to provide reliable flow data. Discharge
measurements are made by continuously recording stage (flow level) in a stilling well
located in each flume or weir structure. Historically, stage was measured with float
gauges and recorded with paper chart recorders. These charts were digitized to create

a continuous record. Stage data were then converted to flow rate with established stage discharge relationships. Currently, each of the 17 watersheds is instrumented with three stage recording devices: 1) a shaft encoder (Sutron Corp, Sterling, VA) connected to a datalogger (Campbell Scientific, Inc., Logan, UT); 2) a float gauge with chart recorder; and 3) an automated sampler with bubbler level recorder (Teledyne-ISCO, Inc., Lincoln, NE). Shaft encoders are the primary stage measurement devices, and the bubblers and float gauges serve as back up devices. Ten of the currently active runoff stations are located at the outlet of small, single landuse watersheds (1.2 to 8.4 ha) to measure "edge of field" processes (Figure 1). Four of the stations are located at the outlet of 0.1 ha plots. Three of the stations are located at the outlet of larger downstream watersheds (17.1 to 125.1 ha) with mixed land uses to evaluate integrated processes.

Historically, sediment loss was the water quality issue of concern at Riesel, but limited nutrient and chemical data were also collected for specific studies. Prior to the 1970's, runoff water samples were collected during runoff events by hand at watershed sites (Knisel and Baird, 1970) and by flow-proportional collectors at field-scale sites. From the 1970's to 2001, runoff water samples were taken with Chickasha samplers (Chichester and Richardson, 1992). In 2001, the mechanical samplers were replaced with electronically operated automated samplers (Teledyne-ISCO, Lincoln, NE). Since 2001, runoff water quality samples have been analyzed for dissolved and sediment-bound nitrogen and phosphorus concentrations and sediment amount.

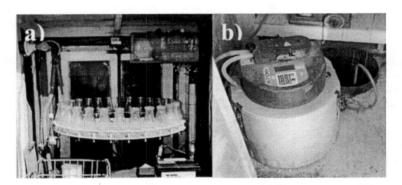

Figure 2: Chickasha samplers used prior to 2001 (a) and electronic automated samplers currently in operation (b).

Rain Gauges
Rainfall data collection began in 1938 and has been conducted at 57 sites throughout the years. Historically, rainfall data were collected by various types of weighing rain gauges and recorded on chart recorders at various intervals during precipitation events. Currently, 15 rain gauges are in operation within 340 ha (Figure 1) making Riesel one of the denser rain gauge networks in the world. At these sites, tipping bucket rain gauges (Hydrologic Services PTY, Ltd., Sydney, Australia) were installed

in the 1990's. Sub-daily rain data are recorded in 10 min intervals (sensitivity 0.254 mm). Throughout the record, a standard rain gauge at each site has been used as a backup and calibration device.

Weather Station
A continuous record of daily maximum and minimum air temperature and precipitation is available for the Riesel watershed since 1938, but additional weather data have been collected during this period. Since 1996, air temperature (average, maximum, minimum), solar radiation, wind speed and direction, precipitation, and maximum and minimum soil temperature have been measured with a weather station (Campbell Scientific, Inc., Logan, UT).

Lateral Flow and Shallow Groundwater Well Stations
A lateral flow station was installed in 1970 to measure lateral subsurface flow from a portion of one watershed. Lateral subsurface flow is intercepted by a series of French drains, collected with a drain pipe, and released into a boxed, sharp-crested v-notch weir, which is used to calculate discharge. Since 2000, twice-weekly discharge from the station has been manually measured. In 1998, seven piezometer wells were installed to monitor shallow groundwater water levels and recharge properties. The water level in these wells is monitored twice-weekly with a hand-held "e-line" water depth gauge.

Telemetry Network
Installation of a radiotelemetry network (Campbell Scientific, Inc., Logan, UT) was completed in 2001. The current network consists of 33 field telemetry stations at 17 runoff stations, 15 rain gauges, and the weather station. The onsite base station communicates with the field stations with a VHF radio signal for equipment maintenance and calibration and realtime monitoring of rainfall, runoff, and weather conditions. An automated data collection schedule runs continuously and collects data daily from each field station. The base station is linked via phone modem to a dedicated computer at the laboratory headquarters in Temple, TX. This phone link also allows remote calibration and automated data transfer.

Historical Contributions and Future Value
When the Riesel watershed was established in the 1930's, little information was available on the impacts of land use and management on runoff, soil loss, and water quality from small agricultural watersheds. Early research at Riesel quantified the ability of a conservation management system (with terraces, grassed waterways, contour farming) to reduce peak flow rates and soil erosion (Baird, 1948; Baird, 1950; Baird, 1964; Baird et al., 1970). Such results contributed the scientific basis for the US conservation farming movement, which is ongoing yet today. More recent research provided fundamental evaluation of the agronomic and environmental effects of tillage, fertilizer, and pesticide alternatives (Baird and Knisel, 1971; Swoboda et al., 1971; Kissel et al., 1976; Richardson et al., 1978; Bovey and Richardson, 1991; Chichester and Richardson, 1992; Richardson and King, 1995; Sharpley, 1995).

The Riesel watershed was also instrumental in the development of watershed models that are now being applied around the world to manage water quality at the field, farm, and basin scale. Results from Riesel have been used to develop routines (Williams et al., 1971) and to calibrate and validate (Arnold and Williams, 1987) the EPIC/APEX (Williams and Sharpley, 1989), GLEAMS (Knisel, 1993), and SWAT (Arnold et al., 1998) models. Subsequent experimentation has also supported model revision and evaluation (Richardson and King, 1995; King et al., 1996; Ramanarayanan et al., 1998; Harmel et al., 2000; Green et al., 2006).

As exemplified by such contributions, the USDA-ARS experimental watershed network, along with long-term research and monitoring sites operated/supported by USGS, Forest Service, National Park Service, and National Science Foundation, are valuable fundamental tools for developing improved understanding of hydrologic processes in various biogeographic regions of the US. Although private, local, and state entities need watershed research results and data, few, if any, have the resources or stated responsibility to conduct long-term integrated watershed research and monitoring (Slaughter and Richardson, 2000). Thus, USDA-ARS and its federal partners have a unique privilege and responsibility to sustain long-term watershed networks to meet the water availability, environmental quality, and food and fiber production demands of the future. The value of these long-term sites should not be disregarded in budget decisions for research and monitoring. Sites with continuous records are particularly valuable for studies designed to identify trends or changes caused by climate shift or other factors and are necessary to determine the influence of extreme, rare events such as floods. The National Research Council (1998) emphasized the need for long-term research and monitoring that is integrated across time and spatial scales and recommended that maintaining sites with exceptionally long-term records should be particularly emphasized (Slaughter, 2000). With 38-71 yr data records from nine geographically diverse locations, the USDA-ARS experimental watershed network fits this profile. The value of these long-term data sets is evidenced by recent publications addressing precipitation (Hanson, 2001; Nichols et al., 2002; Harmel et al., 2003) and discharge and sediment transport (Pierson et al., 2001; Edwards and Owens, 1991; Van Liew and Garbrecht, 2003; Harmel et al., 2006).

Small watershed scale sites such as Riesel are particularly valuable for field- to farm-scale research because of their long-term, detailed, continuous record on multiple small watersheds. Data at that scale are vital to properly evaluate runoff and constituent transport processes from single land use, relatively homogeneous watersheds and to differentiate mechanisms for various land use conditions. Long-term rainfall and flow data are necessary for water supply, water quality, and flood management and for calibration, validation, and application of watershed models. Sediment transport and flow data are needed for the optimal design of culverts, bridges, detention basins, and reservoirs (USDA-SCS, 1942). Data collected at larger scales are often influenced by dams, channel processes, and variable land management, which alter routing and confound land management effects.

Current Research
Research at the Riesel watershed is currently addressing several economically and environmentally important issues. As part of the USDA Conservation Effects Assessment Project, the environmental, agronomic, and economic impacts of poultry litter fertilization over the long-term with repeated annual application are being evaluated (Harmel et al., 2004). Results of this study, which is the most comprehensive field-scale study of its kind, will continue to receive interest from research, regulatory, and legal standpoints because of its integrated approach and the lack of similar data elsewhere.

Although the continuous long-term hydrologic record for Riesel has produced a valuable knowledge of the Texas Blackland Prairie hydrology (Allen et al., 2005; Arnold et al., 2005, Harmel et al., 2003; Harmel et al., 2006), additional research is underway to better understand the mechanism and effects of soil shrink/swell. This Vertisol phenomenon impacts transport of pollutants, maintenance of in-stream flows, and stability of road and building foundations. Other research is examining the effects of changing rainfall patterns due to potential climate change. Rain-exclusion shelters are being used to study effects of altering the timing and quantity of precipitation events on forage production and plant species composition on remnant prairies.

CONCLUSIONS
With sites, such as Riesel that have made many important scientific and socioeconomic contributions, the USDA-ARS experimental watershed network has a unique opportunity and responsibility to provide continuous, watershed-based data to federal, state, and local governments, universities, and private organizations that need this information. The continued support and commitment to this and similar networks, in which the substantial initial investment in infrastructure has already been made, should remain a national priority in spite of budget pressures. It is only with information gathered from such sites that water availability, environmental quality, food and fiber production, and climate change can be addressed on the national scale.

ACKNOWLEDGMENTS
We would like to recognize the efforts of many technicians that have contributed to the success of the Riesel experimental watershed. Successful operation requires a talented and dedicated staff; and we have been blessed in this regard. We especially want to recognize current staff members (Lynn Grote, Steve Grote, James Haug, Gary Hoeft, and Larry Koester) for their outstanding efforts in equipment maintenance, data collection, and record keeping.

REFERENCES
Allen, P.M., R.D. Harmel, J.G. Arnold, B. Plant, J. Yeldermann, and K.W. King. (2005). Field data and flow system response in clay (vertisol) shale terrain, North Central Texas, USA. *Hydrological Processes* 19:2719-2736.

Arnold, J.G. and J.R. Williams. (1987). Validation of SWRRB - Simulation for Water Resources in Rural Basins. *Journal of Water Resources Planning and Management* 113(2):243-256.
Arnold, J.G., R. Srinivasan, R.S. Muttiah, and J.R. Williams. (1998). Large-area hydrologic modeling and assessment: Part I. Model development. *Journal of the American Water Resources Association* 34:73-89.
Arnold, J.G., K.N. Potter, K.W. King, and P.M. Allen. (2005). Estimation of soil cracking and the effect on surface runoff in a Texas Blackland Prairie watershed. *Hydrological Processes* 19:589-603.
Baird, R.W. (1948). Runoff and soil conservation practices. *Agricultural Engineering* May 1948: 216-217.
Baird, R.W. (1950). Rates and Amounts of Runoff for the Blacklands of Texas. USDA Technical Bulletin No. 1022.
Baird, R.W. (1964). Sediment yields from Blackland watersheds. *Transactions of ASAE* 7(4):454-465.
Baird, R.W., C.W. Richardson, and W.G. Knisel. (1970). Effects of Conservation Practices on Storm Runoff in the Texas Blackland Prairie. USDA Tech. Bull. No. 1406, 31 pg.
Baird, R.W. and W.G. Knisel. (1971). Soil Conservation Practices and Crop Production in the Blacklands of Texas. USDA Conservation Research Report No. 15, 23 pg.
Bovey, R.W. and C.W. Richardson. (1991). Dissipation of clopyralid and picloram in soil and seep flow in the Blacklands of Texas. *Journal of Environmental Quality* 20:528-531.
Chichester, F.W. and C.W. Richardson. (1992). Sediment and nutrient loss from clay soils as affected by tillage. *Journal of Environmental Quality* 21(4):587-590.
Edwards, W.M., and L.B. Owens. (1991). Large storm effects on total soil erosion. *Journal of Soil and Water Conservation* 46(1):75-78.
Green C.H., J.G. Arnold, J.R. Williams, R. Haney, and R.D. Harmel. (2006). Soil and Water Assessment Tool hydrologic and water quality evaluation of poultry litter application to watersheds in Texas. *Transactions of ASABE* (In Review).
Hanson, C.L. (2001). Long-term precipitation database, Reynolds Creek Experimental Watershed, Idaho, United States. *Water Resources Research* 37(11):2831-2834.
Harmel, R.D., C.W. Richardson, and K.W. King. (2000). Hydrologic response of a small watershed model to generated precipitation. *Transactions of ASAE* 43(6):1483-1488.
Harmel, R.D., K.W. King, C.W. Richardson, and J.R. Williams. (2003). Long-term precipitation analyses for the central Texas Blackland Prairie. *Transactions of ASAE* 46(5):1381-1388.
Harmel, R.D., H.A. Torber, B.E. Haggard, R. Haney, and M. Dozier. (2004). Water quality impacts of converting to a poultry litter fertilization strategy. *Journal of Environmental Quality* 33:2229-2242.
Harmel, R.D., C.W. Richardson, K.W. King, and P.M. Allen. (2006). Runoff and soil loss relationships for the Texas Blackland Prairies Ecoregion. *Journal of Hydrology* 331:471-483.

King, K. W., C.W. Richardson, and J.R. Williams. (1996). Simulation of sediment and nitrate loss on a vertisol with conservation tillage practices. *Transactions of ASAE* 39(6):2139-2145.

Kissel, D.W., C.W. Richardson, and E. Burnett. (1976). Losses of nitrogen in surface runoff on the Blackland Prairie of Texas. *Journal of Environmental Quality* 5(3):288-293.

Knisel, W.G. and R.W. Baird. (1970). Depth-integrating and dip samplers. *Journal of the Hydraulics Division of ASCE* 96:497-507.

Knisel, W.G. (1993). GLEAMS Groundwater Loading Effects of Agricultural Management Systems, v. 2.10. UGA-CPES-BAED Publication No. 5.

Nichols, M.H., K.G. Renard, and H.B. Osborn. (2002). Precipitation changes from 1956 to 1996 on the Walnut Gulch Experimental Watershed. *Journal of the American Water Resources Association* 38(1):161-172.

Pierson, F.B., C.W. Slaughter, and Z.K. Crane. (2001). Long-term stream discharge and suspended sediment database, Reynolds Creek Experimental Watershed, Idaho, United States. *Water Resources Research* 37(11):2857-2861.

Ramanarayanan, T.S., M.V. Padmanabhan, G.H. Gajanan, and J.R. Williams. (1998). Comparison of simulated and observed runoff and soil loss on three small United States watersheds. Modelling Soil Erosion by Water, J. Boardman and D. Favis-Mortlock (Ed.), NATO ASI Series, Vol 155, pg. 75-88.

Richardson, C.W. and K.W. King. (1995). Erosion and nutrient losses from zero tillage on a clay soil. *Journal of Agricultural Engineering Research* 61:81-86.

Richardson, C.W., J.D. Price, and E. Burnett. (1978). Arsenic concentrations in surface runoff from small watersheds in Texas. *Journal of Environmental Quality* 7(2): 189-192.

Sharpley, A.N. (1995). Identifying sites vulnerable to phosphorus loss in agricultural runoff. *Journal of Environmental Quality* 24:947-951.

Slaughter, C.W. (2000). Long-term water data...wanted? needed? available? *Water Resources IMPACT* 2(4):2-5.

Slaughter, C.W. and C.W. Richardson. (2000). Long-term watershed research in USDA-Agricultural Research Service. *Water Resources IMPACT* 2(4):28-31.

Swoboda, A.R., G.W. Thomas, F.B. Cady, R.W. Baird. (1971). Distribution of DDT and toxaphene in Houston Black Clay on three watersheds. *Environmental Science and Technology* 5(2):141-145.

United States Department of Agriculture Soil Conservation Service. (1942). The Agriculture, Soils, Geology, and Topography of the Blacklands Experimental Watershed, Waco, Texas. Hydrologic Bulletin No. 5. 38 pg.

Van Liew, M.W. and J. Garbrecht. (2003). Hydrologic simulation of the Little Washita River Experimental Watershed using SWAT. *Journal of the American Water Resources Association* 39(2):413-426.

Williams, J. R., E. A. Hiler, and R.W. Baird. (1971). Prediction of sediment yields from small watersheds. *Transactions of ASAE* 14:1157-1162.

Williams, J.R. and A.N. Sharpley. (1989). EPIC − Erosion/productivity impact calculator: 1. Model documentation. Technical Bulletin No. 1768. Washington, D.C.: USDA Agricultural Research Service.

Watershed Research at the North Appalachian Experimental Watershed at Coshocton, Ohio

James V. Bonta, Lloyd B. Owens, Martin J. Shipitalo[1]

[1]USDA-Agricultural Research Service, North Appalachian Experimental Watershed, Box 488, Coshocton, Ohio 43812-0488; PH (740) 545-6349; FAX (740) 545-5125; email: bonta@coshocton.ars.usda.gov

Abstract

The North Appalachian Experimental Watershed (NAEW) at Coshocton, Ohio was established during the mid 1930s as one of the first watershed research locations in the US. The mission of the outdoor laboratory facility is to determine the effects of land-management practices on hydrology and erosion, to investigate scaling from small plots to large watersheds, and to determine rates and amounts of runoff from watersheds of varying configuration, shape, cover, topography, land-management practice. The NAEW infrastructure consists of approximately 1050 acres that includes large lysimeters, small and large experimental watersheds, and a network of rain gauges. One of the first land-management practices investigated was an intensive study on the effects of a crop rotation on steep watersheds with different soils. These early studies contributed to the development of the no-till concept for farming steep lands to reduce erosion and runoff. No-till has been investigated continuously for 43 years at the NAEW with the current emphasis on effects on soil quality, carbon sequestration, and crop residue removal for biofuel production. Data from Coshocton were included in the original development of the curve number method, which is used worldwide. Watershed studies investigations include effects of conservation tillage, herbicide and nutrient management, pasture management, coal mining and reclamation, and urbanization on hydrology and water quality. Other studies conducted throughout the history of the NAEW include those on rain gauges, soil carbon, evapotranspiration, precipitation simulation, ground-water recharge, curve numbers, macropores, hydraulics, watershed modeling, and instrumentation development. Expertise and data at the NAEW are sought after worldwide on these topics. The facility has unique features that enable it to contribute to watershed model component development, including identification and quantification of the processes of preferential flow, interflow, non-uniform runoff generation, and scaling. The NAEW continues to have an important impact on soil and water conservation.

Background

Management of agricultural fields was recognized as a major factor that affected runoff and sediment losses from agricultural areas prior to the 1930s. There were few studies for which the effectiveness of management practices had been evaluated, however, and the science underlying management and other factors had not been successfully incorporated into practical engineering procedures. Adverse weather experiences (droughts and floods) affecting the agricultural community in the U.S. led to a national initiative to develop science-based guidance on management of agricultural areas. In this initiative, an experimental watershed facility was to be established in each of ten regions of the US. The North Appalachian Experimental Watershed (NAEW) near Coshocton, Ohio was established in the late 1930s as part of the larger national plan. The mission of the NAEW is expressed in the following three objectives (Ramser and Krimgold, 1935):

A) "To determine the effect of land use and erosion control practices upon the conservation of water for crops and water supply and upon the control of floods under conditions prevailing at the North Appalachian Region [NAR] of the US";

B) "To determine the effect under (A) for small and large areas and to trace variations in this effect from the smallest plot and lysimeters through a series of intermediate watersheds to the largest watershed on the project"; and

C) "To determine the rates and amounts of run-off for precipitation of different amounts and intensities for watershed typical of the NAR of different configuration, size, shape, topography, cover, underground conditions, land use, and erosion control practices. To furnish data needed for use in the design of erosion control structures and in the design and operation of the Muskingum Watershed Conservancy District and other flood control projects lying within the NAR".

The NAEW was originally operated by the USDA - Soil Conservation Service, Division of Research, and in 1954 became part of the newly created Agricultural Research Service. The purpose of this paper is to review the history of the NAEW, and to present the research portfolio and the unique capabilities of this outdoor laboratory for land and water management research.

Physical Setting

Site Selection and Representativeness: The primary factor for siting the NAEW was its "representativeness" of the North Appalachian Region, and in particular E. Pennsylvania, N. Kentucky, W. Virginia, and southeastern Ohio. "Representativeness" was quantified by the following factors: drainage basin characteristics, topography, geology, major soil types, climate, major type of erosion, and major land-use practices (Ramser and Krimgold, 1935). Available maps of these factors were overlaid and regions of representativeness were identified, narrowing the number of sites to 86. Evaluation of a matrix of site-specific factors led to selection of the NAEW on which many small and large experimental watersheds were planned.

General Characteristics: The NAEW is in east-central Ohio north of Coshocton, Ohio and originally included two land areas, the Little Mill Creek Watershed (LMC; 7.46 mi^2) and the present boundaries of the 1050-ac NAEW facility (Figure 1). The facility is in the uplands west of the northern Appalachian Mountains with slopes typically of the order of 18-25%. Elevations range from 900 to 1150 ft sld. Land use is approximately 15% cropland (corn, soybeans, and wheat), 55% grassland (pasture and hay fields), 26% woodland, and the rest miscellaneous. The NAEW has control of all land, facilitating land management research.

The present-day NAEW is composed of small upland watersheds typically of the order of 1-ac, comprising the infrastructure of small experimental watersheds. These watersheds typically have no well-defined, incised stream channel, but concentrate runoff in swales. The upland watersheds are ephemeral, and run off for relatively short durations under heavy rainfall, snow melt, and wet antecedent soil conditions. Downstream from the uplands, the landscape is dissected by incised stream channels, and drain larger areas up to approximately 300 ac, some of which are gauged.

Climate: The study site is exposed to cold dry air from the northwest and warm moist air from the south, with a humid-temperate continental climate. The average annual precipitation is 959 mm and the average annual temperature is 10.4 °C.

Soils: Three main soil types on the NAEW include residual soils that are well drained and derived from sandstone, soils that are derived from shale and have an argillic horizon, and soils that have characteristics between these two extremes. Principal soil series occupying most of the landscape are Berks, Coshocton, Dekalb, Keene, and Rayne soils.

Geology: The bedrock on the facility lies within the Pennsylvania System (Allegheny Formation) and is composed of unglaciated, sedimentary strata consisting mainly of sandstone and shale with interbedded strata of coal, clay, and limestone (Kelley et al. 1975). Geologic structure is characterized as having an underlying anticline (with local synclines), and by strata dipping to the southeast at an angle of less than 1°. This geologic structure is important to the local hydrology of the NAEW.

Importance of Soil and Geology on Hydrology, Research, and Monitoring Capabilities

The soils and geology of the NAEW are important to the hydrology, the current monitoring infrastructure, and the capabilities of the NAEW to address high priority research needs of the scientific and stakeholder communities.

Some clay layers are spatially persistent and *springs* form at the intersection of the outcrop and the land surface. Also, it has been observed that runoff increases as watershed area increases because as stream channels dissect the landscape from the upland to lowland areas, they intercept water tables perched on the clay and limestone layers, thereby increasing the base flow in streams. This increase in total runoff as watershed area increases is important to help understand and quantify *scaling* of runoff responses in current watershed modeling efforts.

The physical features of the NAEW facility make it an ideal location to study the hydrological processes associated with *nonuniform generation of runoff*, processes that are not well quantified in current watershed models, but that can significantly influence observed runoff in stream channels. This is because the springs cause wet areas on the landscape to generate runoff sooner during a storm than surrounding areas that tend to dry out more quickly between runoff events. Furthermore, soils and land management are spatially diverse, causing runoff to be generated nonuniformly.

The geology and soils of the facility make the NAEW ideal for quantifying the process of *interflow*, another component in which current watershed models are deficient. For example, field observations have shown that, during runoff events, the areas above the outcrop of clay layers tend to infiltrate water while those down slope from the outcrop tend to exfiltrate water. The interflow leaving the soil has a significant impact on runoff generation and water quality.

The clay layers that perch water tables beneath the NAEW are useful for investigating the large-scale effects of land management on subsurface water quality. The *impact of land management activities on ground water* over an area as large as ~40 ac can be observed by monitoring springs at the edge of the clay layers.

A 7-ac area of the NAEW has a shallow, thick clay layer that outcrops along the periphery of an isolated hilltop ("Urban's Knob", UK). Furthermore, the clay layer has a synclinal structure forcing most water to emerge at a gauged single spring. Consequently, all subsurface water originates from precipitation (none comes from adjacent areas that can be contaminated), making UK a *natural lysimeter*. This area has been instrumented with a network of about 40 monitoring wells and piezometers. This hilltop can be used to determine the impact of land management on subsurface water both in the unsaturated and saturated zones. Furthermore, the shallow ground water system allows a "quick" evaluation of the impacts to ground water.

Macropore flow is also a watershed model component that requires further understanding and quantification. There have been extensive field, lysimeter, and laboratory studies on the effects of soil management on macropore development and their subsequent effects on water movement and chemical transport. The results of these studies have been used to modify the macropore flow component of the Root Zone Water Quality Model (RZWQM).

Instrumentation and Data

The LMC watershed was instrumented with a network of recording rain gauges and nested watersheds ranging in size from 97 ac to 4581 ac (Figure 1). The LMC infrastructure was discontinued after about 30 years of watershed data were collected.

The infrastructure of the NAEW consists of small watersheds ranging in size from 0.65 ac to 7.59 ac (Figure 2). Seventeen small experimental watersheds are currently in operation and are instrumented with H flumes for runoff measurement (Figure 3; Brakensiek et al. 1979). Two of the small watersheds have drop-box weirs for expected sediment-laden flows (Bonta and Pierson, 2003). Six larger watersheds are instrumented with short-crested V-notch weirs. Currently, a network of 13 recording rain gauges is in operation along with a meteorological station. Monolith lysimeters ($1/500^{th}$ ac) were installed when the NAEW was established at three sites representing 3 major soil types with 3 or 4 lysimeters at each location . Each lysimeter is ~6 ft wide, ~14 ft long, and ~8 ft deep and one lysimeter at each site is weighed to measure evapotranspiration. All measurements are obtained with data loggers and a radio telemetry system with a time resolution of 1 min.

Samples for water quality analysis from the small watersheds are collected a using Coshocton Wheel sampler (Figure 3). This sampler was invented at Coshocton and collects a constant proportion of runoff water, requires no power, and is used worldwide, particularly in remote areas. At the large watersheds, the Coshocton vane sampler is used (Edwards et al., 1976).

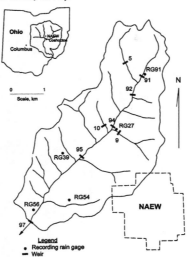

Figure 1. Original land area and Little Mill Creek watershed comprising the NAEW

Runoff, weather, precipitation, lysimeter, and other data have been collected at the NAEW during the last 70 years under widely varying weather and soil and land-management conditions. Consequently, the data have significant value to the hydrological community and have been used for a wide variety of investigations as mentioned in a later section.

Research Projects, Results, and Potential

The objective of the field scale research conducted at the NAEW outdoor laboratory is to develop ways to manage land to improve water quality, reduce flood damage, and to reduce sedimentation, nutrients, and pesticides in water supplies. The NAEW is a resource known for its technical expertise and research results that have been sought out by *many* scientists, universities, farmers and others worldwide. Examples of past and current contributions by Coshocton follow.

Curve number (CN): The CN is a method used worldwide to estimate runoff volumes. Much of the original developmental work on the CN method was conducted at Coshocton using data collected here. Today, data from Coshocton are still used in new CN developments and are used as examples in new documentation.

Long-term no-till/conservation tillage research: Runoff and erosion are greatly reduced and in some cases almost eliminated due to research results from Coshocton. The national farm programs administered by the NRCS incorporate no tillage, a concept largely tested and developed at the Coshocton facility in cooperation with The Ohio State University. This system has several features such as improved soil structure and greater biological activity, reduced fuel/energy needs (i.e., fewer trips across the field), yields equal to or greater than with conventional, plow-disk tillage, yields better during droughts, increased soil carbon storage, especially with manure applications

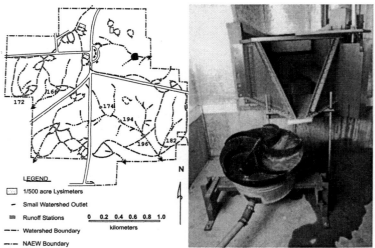

Figure 2. Present-day boundaries and infrastructure of the NAEW.

Figure 3. H flume and Coshocton wheel water sampler.

Pesticide transport: Conservation tillage practices usually rely on herbicides for weed control. NAEW research has shown that concentrations in some herbicides in runoff can reach levels of concern. These concerns can be addressed by reducing application rates by banding or by replacing residual herbicides with short half-life herbicides, such as glyphosate applied to genetically modified glyphosate tolerant crops.

Biofuel-related research: Research at Coshocton has demonstrated that removal of large amounts of crop residue for ethanol production can negate many of the soil quality benefits of long-term no-till.

Crop rotations: An extensive study of the effects of crop rotation and soil type in the early history of the watershed indicated that most soil loss occurred when crow crops were produced and that most of this soil loss was due to a few infrequent severe storms.

Environmental impacts of grazing systems: Environmental recommendations for fertilizer application rates to pastures have been developed. For example, mineral N fertilizer rates in excess of 200 lbs/Ac annually will cause high NO3-N levels in subsurface water; and using grass-legume mixtures for pastures instead of highly fertilizer grass is environmentally sound. The monitoring infrastructure, built on the configuration of clay layers, has allowed a comprehensive analysis of grazing impacts on nutrient loads in surface and subsurface waters.

Management Intensive Grazing (MIG): The frequent rotation of livestock between small paddocks is being evaluated in terms of impacts on surface and subsurface water quality, animal health, and changes in plant species. Benefits of MIG to the producer include extended grazing season, less cost, meat and dairy products with improved human health benefits, and more leisure time.

Environmental impacts of urbanization: A new project that examines increased imperviousness and spatial location of imperviousness on experimental watersheds, as well as best management practices that will reduce runoff volumes and peak flows.

Winter application of manure on frozen soil: Plots have been established to measure impacts on runoff water quality resulting from manure application to frozen ground, following NRCS guidelines.

Pathogens from manure applications: In collaboration with the USEPA, pathogen transport from experimental watersheds and plots having applied beef, turkey, and swine manure will be investigated to help develop science-based guidelines for manure applications by producers.

Instrumentation: The NAEW has developed and adapted many hydrological and water-quality instruments including the original Coshocton wheel water sampler, the Coshocton vane sampler, large particle sampler, drip-flow meter and sampler, hydraulic studies of the drop-box weir, the adaptation of the Coshocton wheel for the drop-box weir, precipitation gauge, natural precipitation infiltrometer, worm-burrow infiltrometer, and rainfall simulator for macropore studies.

Precipitation modeling: Computer generation of short time increment precipitation intensities as they vary with season and location in the US has been developed, and is useful where long precipitation or short-time increment records do not exist. The long-term precipitation records of the NAEW are being used as the primary data source for concept development.

Preferential movement of water in soil: Investigations have been conducted on how preferred pathways through the soil affect runoff and water quality. These include effects of earthworm burrows and cracks on movement of liquid manure to drain tiles, development of fundamental knowledge of the fate of water and chemicals in subsurface water in soil. This work at Coshocton is recognized worldwide.

Evaluation of effectiveness of best management practices (BMPs): A current investigation into methods to quantify the effects of BMPs when there are few data, and to account for natural variation in watershed response due to weather. Water quality data from the Coshocton watersheds are being used in this pilot project in collaboration with the USEPA.

Filter sock performance evaluation in grassed waterways: Coshocton watershed data have documented the limitations of on-field control of pesticides. An investigation is underway to explore whether water treatment residuals/wastes can be used to filter pesticides and nutrients from surface runoff from agricultural land.

Paper mill byproducts: The benefits and disadvantages of using paper mill byproducts as a reclamation aid for re-vegetating surface mined land is a currently being investigated in collaboration with the Ohio state agencies and industry.

Carbon sequestration: An experimental watershed at Coshocton with the longest record of runoff with over 40 years of continuous no-till corn has been the source of much research related to soil carbon sequestration. This and other watersheds have enabled the evaluation of the effects of land management on stored carbon.

Climate change: The long-term data base at Coshocton is useful for evaluating the changes in weather, runoff, and evapotranspiration due to possible changing climate.

Surface mining and reclamation: A landmark comprehensive study on how coal mining and reclamation affects runoff and water quality in surface and ground water was conducted by the NAEW in collaboration with The Ohio State University. The data collected allow the investigation of the very long-term effects of mining.

Other research: Throughout the history of the NAEW, numerous research projects have investigated many other aspects of hydrologic research including rain gauges, soil moisture, evapotranspiration, landfill caps, watershed modeling, ground-water recharge, interflow, natural precipitation infiltration, erosion control, etc.

Summary

Throughout its history, the experimental watershed facility at Coshocton has been addressing the needs of the nation, and is recognized nationally and internationally. The research portfolio at Coshocton covers broad areas of interest to many stakeholders. The strengths of the Coshocton facility are its data base, monitoring infrastructure, physical features, control of land, and expertise. The infrastructure allows investigation of new land management priorities as they arise without additional cost. Furthermore, often the historic data base contains the baseline data needed for land-management evaluations, reducing the time required for research. The physical features of the facility are ideal for addressing the weaknesses in watershed modeling such as interflow, preferential flow, watershed scaling, and non-uniform generation of runoff. The facility is also ideal for educating future scientists on runoff and water quality processes that occur naturally within natural soil variability.

References

Bonta, J.V. and F.B. Pierson. 2003. Design, measurement, and sampling with drop-box weirs. Applied Engineering in Agriculture 19(6):689-700.

Brakensiek, D.L., H.B. Osborn, and W.J. Rawls. 1979. Field Manual for Research in Agricultural Hydrology. U.S. Dept. of Agriculture, Agriculture Handbook No. 224. U.S. Government Printing Office, Washington, DC. 547 pp.

Edwards, W.M., H.E. Frank, T.E. King, and D.R. Gallwitz. 1976. Runoff sampling: Coshocton vane proportional sampler ARS-NC-50.

Kelley, G. E., W. M. Edwards, L. L. Harrold, and J. L. McGuinness. 1975. Soils of the North Appalachian Experimental Watershed. USDA Misc. Publ. #1296, 145 pp.

Ramser, C. E., and D. B. Krimgold, 1935. Detailed working plan for watershed studies in the North Appalachian Region relating to water conservation. flood control, and run-off as influenced by land use and methods of erosion control. WHS#1, November, 1935.

45 Years of Climate And Hydrologic Research Conducted at the Reynolds Creek Experimental Watershed

Gerald N. Flerchinger[1,2], Danny Marks[1], Mark S. Seyfried[1], Fred B. Pierson[1], Anurag Nayak[1], Stuart P. Hardegree[1], Adam H. Winstral[1], and Patrick E. Clark[1]

[1]USDA-ARS Northwest Watershed Research Center, 800 Park Boulevard, Suite 105, Boise, Idaho 83712; PH (208) 422-0700; [2]email: gflerchi@nwrc.ars.usda.gov

Abstract

The Reynolds Creek Experimental Watershed was established over 40 years ago to conduct research and collect data to address critical water issues on western rangelands. Research focus on the watershed has changed over the past 40 years, starting initially with monitoring and describing hydrologic processes and migrating toward development of computer-based tools to address critical water supply, water quality and rangeland management problems. Data quality at the watershed has improved greatly with improved instrumentation but the type of data collected has also changed to meet new research needs. The combination of the long historic records, current data collection and support staff make RCEW a unique "outdoor laboratory" to address critical natural resource questions.

Introduction

Rangelands and the water they yield are under increasing pressures by multiple users. Complex landscapes and variable weather conditions make it difficult to manage natural resources over large landscapes. Landscape disturbance caused by wildfire and invasive weeds complicate water resource and land management issues because they occur over very large areas, which are difficult to study.

The USDA Agricultural Research Service, Northwest Watershed Research Center (NWRC) has been conducting hydrologic and rangeland research at the Reynolds Creek Experimental Watershed (RCEW) since 1960. The research mission at NWRC and RCEW has evolved over the last 45 years, but is still focused on the type of long-term studies that is difficult for either private industry or the university system to conduct. Reynolds Creek watershed (239 km²) is located on rangeland in the north flank of the Owyhee Mountains in southwestern Idaho, USA, approximately 80 km southwest of Boise, Idaho (Figure 1). About 77% of the watershed is under federal or state ownership with the remainder being privately owned. Primary land use of the watershed is livestock grazing with some irrigated fields along the creek at the lower elevations.

Reynolds Creek drains north to the Snake River and ranges in elevation from 1090 m to 2240 m. Extensive hydrologic records have been collected since the experimental watershed was established in the early 1960's. This long-term database and the infrastructure within the Reynolds Creek Experimental Watershed make it a unique setting to address current natural resource concerns including prescribed fire effects, changing climate, and water and rangeland management.

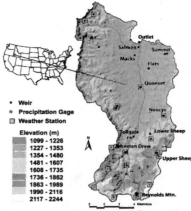

- Weir
- Precipitation Gage
- Weather Station

Elevation (m)
1099 - 1226
1227 - 1353
1354 - 1480
1481 - 1607
1608 - 1735
1736 - 1862
1863 - 1989
1990 - 2116
2117 - 2244

Figure 1: Location map of Reynolds Creek Experimental Watershed showing distribution of weather stations, precipitation sites, and weirs in operation.

Background

In 1958, an increasing awareness of water supply, flooding and sedimentation problems prompted the U.S. Congress to initiate research on regional soil and water conservation problems. In January of 1959, the Secretary of Agriculture appointed a working group to assess facility needs for soil and water research. This working group identified the need for six major experimental watersheds where critical regional soil and water conservation problems could be studied (Senate Document No. 59, 86th Congress). Later that year, Congress appropriated funds for the Agricultural Research Service to initiate research on these sites, one of which was to be located in the Pacific Northwest. Research at the Northwest Watershed Research Center (NWRC) was to focus on a rangeland watershed in a geographic area where the primary precipitation input was snow. The watershed selected for this research was to have the following characteristics: (1) suitable stream channels for streamflow measurements; (2) reasonable range in land use and water production; (3) precipitation in the range of 150 to 750 mm dominated by wintertime precipitation typical of the Pacific Northwest Intermountain region; (4) accessibility and road network within the watershed; (5) possibility of isolating tributary streams; (6) complete range in aspect of unit sources and tributaries; (7) willingness of landowners to cooperate; (8) range in soils and geologies; and (9) appropriate size of 65 to 260 km². After analysis of over 100 watersheds, Reynolds Creek was chosen in 1960 as meeting the above criteria and best representing the diverse soils, geology, topography, vegetation, land use and climate of semi-arid rangelands in the Pacific Northwest. Within that same year, the NWRC began installing precipitation and streamflow instrumentation and conducting resource inventories.

Watershed Characterization

An impressive diversity of climate, soils and plant communities is represented on the watershed. A detailed survey of plant communities was conducted between 1963 and 1965 by botanists on staff at the NWRC. Plant communities typical of the Great Basin are found at the lower elevations while alpine communities occur at the highest elevations. The Soil Conservation Service, U.S. Department of Agriculture, was contracted to do a soil survey of the watershed, which they completed in 1966. Soils derived from granitic, volcanic and lake sediments are present on the watershed. The diverse climate, soils and vegetation found on the watershed range from semi-arid sagebrush rangeland on shallow desert soils receiving 236 mm of precipitation to aspen and Douglas fir stands on deep organic soils that receive 1118 mm of precipitation, 76% of which falls as snow.

Three main weather stations were established to characterize the climate and have been in operation on the watershed since the early 1960's: Quonset, Lower Sheep and Reynolds Mtn., representing low, middle and high elevation areas on the watershed (Figure 1). Precipitation was originally measured at 83 sites, approximately 1 gauge per square mile (2.6 km^2), but this was pared down to approximately 20 representative sites in 1979 (Figure 1).

The result of this early characterization work was the establishment of a scientific infrastructure at Reynolds Creek that included fundamental vegetation, soil, and geologic information as well as long-term hydrological data that have become increasingly valuable for using Reynolds Creek as an "outdoor laboratory" (Robins et al., 1965) to conduct research.

Past Research and Accomplishments

Early Accomplishments

The research focus at RCEW has gone through some significant changes since 1960. Early efforts were focused in two areas: 1) characterizing the gross climate gradients, streamflow and erosion associated with elevation changes, which were poorly understood; and 2) developing methods to accurately and reliably measure weather and streamflow in remote areas under harsh conditions.

NWRC has made some significant improvements in the methodology for measuring weather variables and streamflow. Precipitation gauges of all types underestimate the amount of precipitation that actually falls on the ground. This measurement error increases under windy conditions and when the precipitation falls as snow rather than rain because wind tends to blow the precipitation away from the gauge opening. To compensate for the "undercatch" of the precipitation gauges, researchers at the NWRC developed the dual-gauge precipitation measurement system, which consists of a shielded and an unshielded gauge. The undercatch of the gauge without the wind shield will be more severe than that of the shielded gauge. The actual amount of precipitation that falls on the ground can be estimated from a logarithmic function of the ratio of the shielded and unshielded gauges (Hanson et al., 2004).

Accurate streamflow measurements are critical, but no structure had been developed that would adequately measure streamflow in the steep gravel-boulder channels which pass large amounts of suspended sediment, large rocks and debris. The initial weir constructed at the Outlet site causes sediment and rocks to settle out upstream which have to be periodically excavated for the weir to function properly. To correct this problem, NWRC cooperated with the Albrook Hydraulics Laboratory of Washington State University to produce the SCOV (Self-Cleaning Overflow V-notch) or "drop-box" weir (Figure 2; Johnson et al., 1966). The box immediately upstream from the V-notch is designed to provide sufficient turbulence to pass sediment and rocks and prevent buildup of sediment behind the weir. Drop-box weirs with capacities ranging from 0.5 to 200 m^3/s were subsequently installed at Salmon, Macks, Summit, Lower Sheep, Johnston Draw, and Tollgate sites (Figure 1). This same weir design has since been used to measure streamflow at several sites across the nation and around the world.

Snow has been a primary focus of research at RCEW since its inception. Much of this work has focused on characterization of snow accumulation, drifting, and melt processes. Additionally, the NWRC has worked closely with the Natural Resource Conservation Service (NRCS) to develop and test instrumentation used to measure snow and weather conditions at the SNOTEL sites across much of the Western United States. Because of the emphasis on collecting high quality data, RCEW

Figure 2: A drop-box weir at Tollgate measuring streamflow in Reynolds Creek. Capacity of the weir is 200 m³/s with top opening about 6 m.

has been a location where scientific theory, in the form of mathematical models that predict snowmelt and streamflow, meets the reality of conditions "in the field". As a result, some the first of the hydrologic computer models were developed and tested using data collected at RCEW (e.g. Schreiber et al., 1972, among others).

Recent Developments
    The original infrastructure at RCEW has been greatly extended and modernized in the last decade. Early weather and streamflow observations were collected on spring-driven chart recorders that had to be changed daily or weekly and read by hand. Digital technology has greatly improved the reliability of the data and reduced man-power required to collect the data. Currently, data are transmitted daily, by telemetry, to the NWRC office in Boise. In addition, RCEW continues to be a site for improving methods for measuring snow, soil water, vegetation, soil erosion and livestock grazing.
    Research and measurements at RCEW since the late 1980's have focused less on simple characterization of climate distribution, and more on development of computer-based hydrological models and application of these models to address management-related issues. Model development has progressed hand-in-hand with expansion of data collection efforts at RCEW as data are required to develop and test computer models. Some of these models include:

ISNOBAL Snow Model: The snow energy and mass balance model, ISNOBAL, predicts snow accumulation, redistribution and melt over the land surface. It has been extensively tested and used at RCEW and in the western United States and Canada to simulate snow accumulation and melt. Scientists at NWRC are working with the Natural Resource Conservation Service to incorporate such models into operational water-supply-forecast models for the Western United States.

ERMiT Erosion Model: NWRC scientists worked together with the US Forest Service Rocky Mountain Research Station to develop the Erosion Risk Management Tool

(ERMiT) to assist the US Forest Service in predicting surface runoff and soil erosion rates that result from both wild and prescribed fires in western ecosystems. This research is now helping the Interagency Burned Area Emergency Response teams evaluate erosion risk and recommend strategies to protect the environment, streams, human life and property after wildfires.

The SHAW Model: The Simultaneous Heat and Water (SHAW) model predicts temperature and water at the soil surface and through the soil profile and includes the effects of plant cover and snow.   Information from the model can be used to assess management and climate effects on biological and hydrological processes, including seedling germination, plant establishment, insect populations, soil freezing, infiltration, runoff, and ground-water seepage. The SHAW model has been used worldwide by engineering consultants and researchers for a variety of applications, including: landfill cover design; post-fire hydrologic and streamflow response; snowmelt runoff; microclimate wildlife habitat (such as insects); ground-water recharge; ground-water contamination; seedling germination and weed competition; and frost depth estimation for construction design.

Historic Data and Climate Trends

Precipitation

Average annual precipitation varies greatly with elevation within RCEW and ranges from 236 inches to 1118 inches. Annual precipitation for the Quonset and a high elevation site are plotted in Figure 3, showing cycles between wet and dry periods. Most of the major winter storms move onto the RCEW from the west and southwest, and thus there is more precipitation along the south and southwest sides of the watershed.

Monthly precipitation plotted in Figure 4 demonstrates how precipitation varies by month with elevation between the low elevation Quonset site and a high elevation site. July is the driest month at all sites. Interestingly, June is one of the wettest months at the low elevation site. Months with the highest precipitation at the Quonset are June, November, December, and January each receiving around 30 mm. Highest monthly precipitation at the upper elevations is close to 180 mm during January. Generally, July, August, and September are the driest months on the RCEW, and November, December, and January are the wettest.

Streamflow

Most of the streamflow throughout RCEW comes from spring snowmelt, particularly at higher elevations, as shown for several locations along Reynolds Creek in Figure 5. Average monthly streamflow is greatest during May at all elevations. Lower elevations experience more rain-on-snow events during the winter months, which extends the runoff season from January to June at the lower elevations; streamflow at the highest elevations is largely limited to April, May and June (Figure 5).

The ten highest streamflow events measured in Reynolds Creek are listed in Table 1. The winter flood in December 1964 was the highest peak flow (109 m³/s with a water depth of 4.48 m at Outlet weir) and resulted from rain on snow with frozen soil. Most

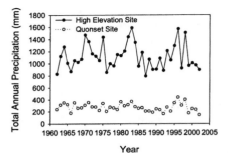

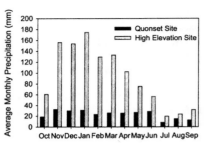

Figure 3: Annual precipitation for the Quonset and a high elevation site on the Reynolds Creek Experimental Watershed.

Figure 4: Average monthly precipitation for the Quonset and a high elevation site on the Reynolds Creek Experimental Watershed.

Table 1. Top ten largest streamflows measured in Reynolds Creek at Outlet weir

Date	Stage Height (m)	Streamflow (m^3/s)
12/23/1964	4.28	109.0
1/13/1963	4.21	63.4
2/15/1982	4.15	59.0
1/111979	3.87	47.1
6/11/1977	3.45	31.7
1/28/1965	3.42	31.5
1/01/1997	3.42	31.1
1/21/1969	3.17	25.1
4/11/1982	3.14	24.4
1/27/1970	2.62	20.6
Average flow	0.52	0.57

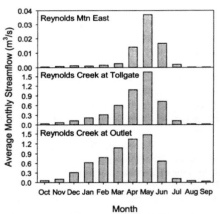

Figure 5: Average monthly streamflow for the three weirs on the main stem of Reynolds Creek.

of the largest runoff events and much of the sediment movement in the stream occur around January and are associated with rain or snowmelt on frozen soil. When the soil is frozen, very little water enters the soil when rain or snowmelt occurs, resulting in large amounts of runoff.

Most frozen-soil runoff events occur in a mid-elevation band on the watershed extending from approximately 1370 to 1740 m. Above this elevation band, the snow is usually deep enough to insulate the soil and keep it from freezing. At lower elevations where the soils typically freeze the deepest, there is typically not enough snow to produce large flooding events. However, when lower elevations do contribute to wintertime flooding, the extent of the floods can be widespread; much of the Pacific Northwest experienced flooding during the 1964, 1969, 1982 and 1997 events listed in Table 1.

Occasional intense thunderstorms tend to be somewhat isolated but can produce significant flooding and erosion. Thunderstorms can occur at any elevation but tend to produce more runoff and erosion at lower elevations where there is typically less vegetation to reduce runoff. The fifth largest peak flow recorded at Outlet (31.7 m³/s) was the result of a short-duration high-intensity thunderstorm centered over Summit Wash in the relatively low-elevation northeastern section of RCEW.

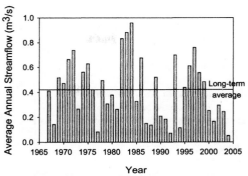

Figure 6: Total annual streamflow measured at Tollgate weir.

Streamflow is quite variable between years as shown for the Tollgate weir in Figure 6. Streamflow during most years is either very high or very low, with very few years considered "average". During high-flow years, streamflow follows the expected pattern of increasing with drainage area: Outlet has the greatest streamflow, followed by Tollgate then by Reynolds Mountain East.

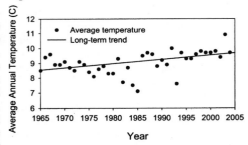

Figure 7: Average annual temperature at Quonset long-term weather station.

During low-flow years, streamflow is often highest at Tollgate because of water being diverted from Reynolds Creek below Tollgate for local irrigation of pasture and hay. The amount of streamflow diverted for irrigation averages around 25% and can vary from 15% during high-flow years to 75% during low-flow years.

Climate Trends

The plot of annual air temperature in Figure 7 shows a fair amount of variation from year to year with the early 1980's and 1993 standing out as colder-than-average years. Even with the yearly variation, the plot suggests a gradual warming trend over the period of record. This slight warming trend can be found at all three long-term weather stations on the watershed. Much of this warming trend can be attributed to warmer springtime temperatures, particularly during March and April.

The trend in warmer springtime temperatures has an interesting effect on streamflow. While there is no detectable change in the total annual streamflow, it appears that more of the streamflow is occurring in the early spring and less in the late spring and summer. Figure 8 presents trends in monthly streamflow for April and June as a percentage of annual streamflow. The percentage of the annual streamflow that occurs in April appears to be increasing over time while the percentage that occurs in

June is decreasing over time. The months of July and August show a decreasing trend as well. This has important implications for irrigation and water supply in the western U.S.

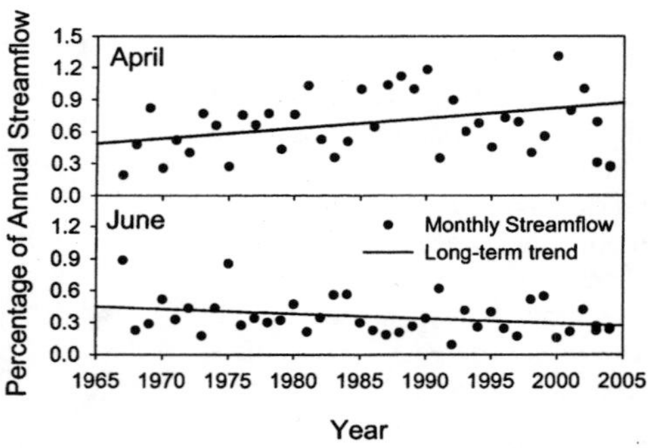

Current Research Directions

At present at RCEW, we have a state-of-the-art "outdoor laboratory" for investigating natural resource questions that facilitates cooperation with scientists nationally and internationally who are able to use the infrastructure available at Reynolds Creek. This infrastructure includes the historic records, the current data collection, and the support and scientific staff at RCEW. Some research programs currently being conducted by NWRC within and outside RCEW are described below.

Figure 8: Percentage of total annual streamflow that occurs in April and June.

Juniper Hydrology and Prescribed Fire Effects

Juniper trees, once mostly isolated to rocky ridges, have considerably expanded their domain to include millions of hectares of rangeland across the western U.S. Uniform stands of dense juniper result in a loss of under-story grasses and shrubs that are critical to wildlife, livestock, and erosion control. Research results have shown a dramatic reduction in erosion after juniper removal and recovery of grasses. However, the effect that juniper has had on the hydrology and streamflow on these lands is uncertain. Prescribed-fire and mechanical treatments are now being used to return juniper to its natural state. NWRC is studying pre- and post-fire hydrology at RCEW to study the hydrologic impacts of juniper and juniper-removal, and also to evaluate prescribed-fire impacts on other vegetation, soil and animal resources.

Snowmelt Models for Streamflow Forecasting

Models developed by NWRC are cutting edge, state-of-the-science tools, but considerable work is required to transform these models into operational tools that the Natural Resource Conservation Service and other action agencies can use. NWRC is working with NRCS to use the snowmelt modeling tools developed at NWRC to improve streamflow forecasting in the Western United States. These streamflow forecasts are critical planning tools used by irrigators, reservoir operators, and municipalities for water supply management.

Livestock Grazing Management

Livestock production is a primary industry on western rangelands. Recently, the NWRC has begun to focus research on livestock grazing management. There are still many unanswered questions regarding the effects of livestock grazing on rangeland environments. These questions have been at the heart of serious land-use debates throughout the West. At the RCEW, range cattle behavior is being evaluated using telemetry tracking collars to determine how prescribed fire treatments, applied for juniper and brush control, affect cattle distribution and activity patterns. A question of particular interest is whether fire treatments alter cattle grazing patterns over the landscape.

Summary

The Reynolds Creek Experimental Watershed was established over 40 years ago to conduct research and collect data to address critical water issues on western rangelands. Research focus on the watershed has changed over the past 40 years, starting initially with monitoring and describing hydrologic processes and migrating toward development of computer-based tools to address critical water supply, water quality and rangeland management problems. Data quality at the watershed has improved greatly with improved instrumentation but the type of data collected has also changed to meet new research needs. The combination of the long historic records, current data collection and support staff make RCEW a unique "outdoor laboratory" (Robins et al., 1965) to address critical natural resource questions.

Data collected over the past 40 years suggest that there is a trend toward warmer springtime temperatures and that snowmelt runoff is occurring earlier in the year than it has historically. Data analysis suggests that more of the annual streamflow is occurring earlier in the year and less during the summer months.

In 1996, NWRC summarized data collection at RCEW and published the data in a series of articles in Water Resources Research (Slaughter et al., 2000). These data are also available over the internet at the anonymous ftp site ftp.nwrc.ars.usda.gov in the directory "database". NWRC is in the process of updating this dataset to extend this publicly-available dataset through the 2005 Hydrologic Year.

References

Hanson, C.L., F.B. Pierson and G.L. Johnson. 2004. Dual-gauge system for measuring precipitation: Historical development and use. Journal of Hydrological Engineering 9(5):350-359

Johnson, C.W., H.D. Copp and E.R. Tinney. 1966. Drop-box weir for sediment laden flow measurement. ASCE Proceedings, Journal of Hydraulics Division 92(HY5):165-190.

Robins, J.S., L.L. Kelly and W.R. Hamon. 1965. Reynolds Creek in Southwest Idaho: An outdoor hydrologic laboratory. Water Resources. Research 1:407-413.

Schreiber, D.L., R.W.Jeppson, G.R. Stephenson, C.W. Johnson, C.M. Cox, and G.A. Schumaker. 1972. Solution of a two-dimensional steady-state watershed flow system, Part II: Evaluation by Field Data. Trans. of ASAE. 15(3):464-473.

Slaughter, C.W. D. Marks, G.N. Flerchinger, S.S. Van Vactor and M. Burgess. 2001. 35 Years of Research Data Collection at the Reynolds Creek Experimental Watershed, Idaho, USA. Water Resources Research, 37(11):2825-2830.

The Galveston/Texas Hurricane of 1900: A Review of the Events that Led to the Galveston Seawall and Grade Raising

by Martha F. Juch, P.E., CFM[1] and Jerry R. Rogers, Ph.D., P.E., D. WRE, F.ASCE[2]

[1] Principal Engineer, CivilTech Engineering, Inc., 600 Round Rock West, Suite 502, Round Rock, Texas, 78681 and Ph.D. student: Texas A & M University.
[2] University of Houston, Department of Civil & Environmental Engineering, Houston, Texas 77204-4003 (jerryrogers@houston.rr.com) (PH: 713-743-4276)

Abstract "At the turn of the twentieth century, the City of Galveston was emerging as one of the most important cities in Texas. The City was the home to the third largest port in the country and was the second-most-heavily-traversed entry for immigrants from Europe. Then on September 8, 1900, the island was virtually wiped out by a hurricane (now classified as a Class 4). With approximately 8,000 to 10,000 fatalities, the event ranks as the deadliest natural disaster to hit the United States. In the wake of the destruction, Alfred Noble (who became ASCE's president in 1903) and Henry Robert (who authored Robert's Rules of Order) were hired as consultants. The engineering recommendation was to raise the structures on the island using 17 feet of hydraulic fill and to construct (on timber piles) a long concrete seawall. Railroad tracks were constructed along the wall to transport the extensive materials needed. All of the utilities for Galveston residents and businesses had to be relocated and raised. In 2001, ASCE recognized the contributions of the Galveston Seawall and Grade Raising and the Corps of Engineers with a ceremony and plaque for a National Historic Civil Engineering Landmark (NHCEL). The account of the storm provided by U.S. Weather Bureau's meteorologist Isaac Cline in the bestseller "Isaac's Storm" are summarized. Engineering aspects of the grade raising and seawall are reviewed."

Talk about "déjà vu". We submitted the idea for this paper in the weeks prior to Hurricane Katrina and later by Rita. Little did we know how timely the subject would be. Needless to say, the years will spur countless comparisons between Galveston and similar cities/ ports along the Gulf Coast and oceans impacted by hurricanes. This paper was written from a plenary session slide presentation made for the Annual Conference of Texas Flood Plain Managers in the fall 2006.

Introduction

In 1900, the city and port of Galveston were located on a barrier island two miles off the Texas coast. Galveston Island is 0.5 to 3 miles wide and 28 miles long. In 1900, the highest land elevation was 8.7 feet. Galveston was the third largest U. S. port and one of the largest Texas cities with 38,000 people. The port of Galveston was the second most heavily traversed U.S. entry for immigrants from Europe.

Hurricane of 1900

Cuba had reported storm conditions and a possible hurricane in late August 1900 but there was little warning for the people in Galveston when the hurricane arrived on September 8, 1900. Galveston had winds of at least 120 mph (hurricane category 4 or 5) and a storm surge of 15 feet covered Galveston Island (at elevations below 8.7 feet). The eye of the 1900 Hurricane crossed the Texas coast between Freeport and Galveston. Between 8,000 and 10,000 fatalities occurred on the island and mainland, which was the deadliest U.S. natural disaster. Around 3600 homes were destroyed, about half of the Galveston homes. Property damage was over $30,000,000, equivalent to $500,000,000 in today's dollars.

Notes from Issac's Storm

Issac Cline was the U.S. Weather Bureau's meteorologist in Galveston. This book was written using Isaac Cline's telegrams, letters, and reports and testimony from survivors and present scientific knowledge about hurricanes. Issac Cline did not recognize early signs and the Cuba information about the approaching hurricane. The U.S. Weather Bureau did not respect Cuban meteorologists and had a ban on Cubans sending weather- related telegraphs. Aside from the roaring surf and rising swales, the Galveston weather the morning of September 8, 1900 included only light rain and escalating winds. In his autobiography, Cline stated he warned Galveston people of the impending disaster; however, in *Issac's Storm*, the author concludes that no such warnings were given.

Emergency Relief Efforts

The National Guard was mobilized to keep order during debris and corpse removal. A local relief committee convened quickly to distribute food and clean drinking water. All able-bodied men were required to help search for and removal of the dead. Photographs from the Galveston County Historical Museum showed wagons and a barge to transport dead bodies and a relief train with food, water, medical supplies and other provisions from all over the nation, which initially arrived in Galveston by boat. On September 21, train service resumed over a new bridge, paid for by all railroads. Mail, telegraph and basic water services were restored within five days. A week after the storm, the local newspaper resumed publication and electrical service was restored to some commercial buildings. Red Cross founder Clara Barton arrived in Galveston Sept. 17, and local doctors are credited with preventing epidemics.

Robert Commission Formed, Recommends a Concrete Seawall

The Galveston City Commission and Galveston County Commissioners' Court appointed a three consulting engineers board to devise plans for protecting the city from future hurricanes.
The engineers' board was to make recommendations regarding:
- Protecting the city from overflows
- Raising the city above overflows, and
- Building a seawall.

The Robert Commission included General Henry Martyn Robert, Alfred Noble and Henry Clay Ripley. General Robert was former Chief Engineer of the Army Corps of Engineers and the Corps' Southwestern Division Engineer and was author of the parliamentary procedures document: ROBERT'S RULES OF ORDER. Alfred Noble was one of the most widely recognized civil engineering consultants, became ASCE National President in 1903, and helped determine the route of the Panama Canal. Henry Clay Ripley lived on Galveston Island, was a civil engineer with the Corps Galveston District, and surveyed the Galveston gabion jetties in 1874 and part of the Houston Ship Channel.

In 1902, the Robert Commission recommended the following:

-Construction of a Galveston seawall with a curved seawall face to deflect upward incoming waves, a seawall height of 17 feet above mean low tide, and 3 miles in length.

- Hydraulic fill pumped behind the seawall with a crest one foot higher than the seawall and 200 feet behind the seawall.

- Brick pavement 35 feet back from the seawall to protect the fill surface with soil and Bermuda grass 60 feet back from the seawall.

The estimated project cost was $3,505,040.

The initial Galveston seawall completed in 1904 is considered one of the great engineering feats of the 20th century. The initial Galveston seawall was 17,593 feet long, with a concrete section 16 feet wide at the base and 5 feet wide at the top with a curved gulf-side face to deflect waves. The weight was 40,000 pounds per foot. The concrete seawall had timber piles driven underneath and was protected by riprap of 4ft. x 4 ft. granite blocks extending 27 feet outward from the toe of the concrete seawall. The 100 foot wide embankment consisted of a 16 feet wide sidewalk, a 54 feet wide brick driveway, and 30 feet of sod and soil. Galveston hosted a large party when the initial Galveston Seawall was completed in 1904.

Galveston Island Grade-Raising

Construction crews lifted more than 2100 of the Galveston's remaining buildings to pump sand underneath them. These included homes, multi-ton churches, mansions, and commercial buildings. The fill area encompassed 500 city blocks. Canals were dug through the city for dredge boats to bring 16.3 million cubic yards of fill material to raise Galveston Island 16-17 feet. Cost of the total project was $6,000,000 (about $122.5 million in today's dollars).

Seawall Extensions

Among the Galveston Seawall extensions after the initial 1902-1904 seawall construction were:

- 4935 ft. to protect the Ft. Crockett Military Reservation (1904-05)
- 10300 ft. to protect the harbor facilities (1918-21)
- 2800 ft. east (1923-26)
- 2800 ft. west (1926-27)
- 1 mile west built by Galveston County (1953)
- 2 miles west built by the Corps of Engineers (1958-63)

The present length of the seawall is 54,790 feet or 10.04 miles of seawall.

Summary and Conclusions

The hurricane of 1900 on September 8 was the deadliest natural U.S. disaster with 8000-10000 dead and half the Galveston buildings destroyed. The Robert Commission of civil engineers: General Henry Martyn Robert, Alfred Noble, and Henry Clay Ripley recommended the 3-mile Galveston seawall construction with a railroad section to transmit the huge material quantities for the curved, Gulf-side concrete seawall on timber piles and rip-rap. The Galveston Seawall and Grade-Raising was a massive civil engineering project requiring excavated canals to allow dredges to convey large quantities of fill material to raise the island to 16 to 17 feet (more than 7-9 feet of fill). All utilities and all buildings had to be raised for the new heights. On October 13, 2001 at the ASCE Annual National Convention, ASCE recognized the contributions of the Galveston Seawall and Grade Raising and those of the Corps of Engineers with a ceremony and plaque for a National Historic Civil Engineering Landmark (NHCEL) located along the Galveston Seawall and Seawall Boulevard. The 2005 Gulf-Coast Hurricanes of Katrina and Rita remind all of the severity of hurricanes which was the case for the great 1900 Galveston hurricane.

References

Bartee, Clark 2001."Galveston Seawall & Grade Raising Project," *INTERNATIONAL ENGINEERING HISTORY AND HERITAGE*, ASCE, pp.461-469.

"Galveston's Bulwark Against the Sea- History of the Galveston Seawall" 1981. U.S. Army Corps of Engineers- Galveston District- Public Affairs Office.

Galveston County Historical Museum 2005. Historical Photographs, Christine S. Carl.

"Galveston Seawall and Grade Raising" 2000. ASCE National Historic Civil Engineering Landmark Nomination, Allen Beene, Texas Section ASCE History Committee.

"Hurricane Resource Center Webpage 2005. www.galvestonhistory.org/hrc-fact- sheet.html, Galveston Historical Foundation.

Juch, Martha and J. Rogers 2007. "Summary and Pictures of the 1900 Galveston Hurricane Destruction and Engineering the Galveston Seawall and Grade- Raising," EWRI Congress Proceedings-CD, ASCE/EWRI.

Larson, Erik 2000. *Isaac's Storm*, Vintage Books.

Selected other publications:

Hebert, Paul J., Jarrell, J. D. & Max Mayfield 1993. "The Deadliest, Costliest and Most Intense United States Hurricanes of this Century," NOAA Technical Memorandum NWS NHC, 41 pp.

Rozelle, Ron 2000. *The Windows of Heaven: A Novel of Galveston's Great Storm of 1900* Texas Review Press, Huntsville, TX.

Hydrologic Detectives: Re-creating the July 1938 Flood for HEC-HMS Model Calibration

Melinda Luna, PE[1] and Andrew Ickert, PE[2]

1. Sr. Engineer, LCRA, PO Box 220, Austin, Texas 78767-0220, 1-800-776-5272 ext. 2396, Melinda.luna@lcra.org.
2. Engineer, Halff Associates, Inc., 4000 Fossil Creek Boulevard, Fort Worth, Texas, 76137, 817-847-1422, aickert@halff.com.

This paper was presented at the Oct. 13, 2006 ASCE Texas Section Fall Meeting.

Abstract

The July 1938 flood was a rain event that had approximately 30" of rainfall in less than a 10-day period. Centered over Brady Creek and the upper reaches of the Colorado Basin, the floods caused Buchanan Dam to open 22 of its 37 gates and flood areas such as downtown Austin, Bastrop, and Wharton. To help calibrate hydrologic models for the Colorado Basin for an LCRA/Army Corps of Engineers study, limited 1938 rainfall and stream flow data were collected from the Corps of Engineers and USGS for the 1938 flood and converted into electronic format. The rainfall data was used to produce Mean Areal Precipitation using the Corps of Engineers HEC-PRECIP program. The PRECIP output data was used as input to the 21,745 square mile HEC-HMS model of the Colorado River basin. This paper documents the source of the data used, the QA/QC methods used for the input and output of the data, comparisons of synthetic data results to observed data and the limitations and solutions to the various methods used to create the data needed for the hydrologic runs. The methodology of the process of re-creating electronic data to test and help calibrate hydrologic models is documented in the paper, as well as lessons learned about the response of river basins to large events such as the 1938 flood. This project also serves as an example of how history can help civil engineers working on hydrologic and hydraulic models.

Background

Initially the re-creation was made with a half dozen hourly gages and the daily rainfall information provided in the 1944 *USGS Water Supply Paper 914* and a memo from the Corps of Engineers dated August 26, 1938. After this initial run was complete, a set of notes was found at the Corps of Engineers from a study done in August of 1944 to create an isohyetal map of the 1938 storm. This data contained additional hourly data and 6-hour data. After review of the data, it was decided that this shorter time-step would produce a better distribution of rainfall than that of the daily values. An initial run with one-third of the data was made to see if a better distribution of rainfall resulted. The remainder of the shorter time-step data was then added to create a final run. For the final run, approximately 75% of the data used was in 6-hour increments whereas the first run was mostly daily rainfall.

1938 Rainfall Data

The daily rainfall information was taken from the 1944 *USGS Water Supply Paper 914* "Texas Floods of 1938 and 1939". Daily rainfall was taken from page 25 Table 5 of this USGS report. To separate trace amounts of rainfall from zero rainfall, 0.005 inches was used to represent trace measurements. The report states that Trace or T is less than 0.01 inch. There were rainfall amounts missing from this report. These locations were Art, El Dorado, and Winchell. These data points were filled by using adjacent rainfall amounts and considering the total rainfall. This data was used in the initial trial. Only 13 of these gages listed in the USGS report were used in the final run. Other daily rainfall data was found in the Corps of Engineers files and utilized. One additional gage not listed in the USGS report was used in the final run (Texon).

Some hourly data included in this rainfall was taken from a Corps of Engineers memo dated August 26, 1938. This data collected rainfall in the Brady Creek area. This data was also taken and converted to an electronic format. Additional hourly gage data was found at the Fort Worth Corps office from a study completed in August 1944.

For the rainfall observers, there were two sources of data, Table 6 in the USGS report and the memo from the Corps dated August 26, 1938. These data sources presented total rainfall for July 16 to 25, 1938. These figures were used as miscellaneous gages in the PRECIP run. These storm totals were added to the miscellaneous gages.

The observer list from the Corps data was compared to the USGS Table 6 so no overlap of information was entered. In the 6-hour data the rainfall observers were labeled U.S.E.D. for United States Engineering Department. The August 1944 study contained this 6-hour data that was used in the final run. The shortest time-step was used in the final run making sure that there were no overlaps in data. In other words, if there were daily, 6-hour and hourly data for one location, only the hourly was used. Only the data that included a location by latitude and longitude was utilized, or the latitude and longitude were estimated by finding the small town with the same rainfall gage name. All gages used were within the Colorado River basin or approximately 40 miles from the Colorado River basin.

The final run in PRECIP was made using 33 sites with storm totals, 14 daily, 172 6-hour and 12 1-hour (hourly) rainfall gages for a total of 231 gages. A table was created that presented the rainfall data used in the final run. It is sorted by the time step of the data and then alphabetical order. The table indicated the time step of the data, name of the location, latitude and longitude, if the data was cumulative or incremental and the source of the data. It is not included in this version of the paper due to its large size and limitation on page length.

QA/QC of Rainfall Data

Several methods were used to check the data being placed into electronic HEC-DSS format. Since the 6-hour data was entered directly into DSS, DSS was utilized to check the data. The 6- hour data was compared to the daily data. The rainfall data for the same location was plotted to check that the overall total and daily totals matched.

The total rainfall was also checked visually by creating what was termed a pseudo grid. This grid was then compared to the isohyetal maps from the USGS report. If a point was found that seemed out of place, the data was reviewed. A visualization of the rainfall termed a pseudo grid was created for each day with a total for the 10 days and also a version of this visualization included a cumulative amount day by day. These visualizations helped show how the storm progressed day by day.   The comparison of the isohyetal maps from the USGS report with the pseudo grid created in this project for the 10 days of rainfall matched reasonably well.   Figure 3 shows the density of rainfall gages and the USGS isohyetal maps.

To check latitude and longitude a shape file was created using the data. Each point was checked to make sure that the description of the location was in the area of the description. Typos were found using this method and the locations corrected.

PRECIP Program: Method Used to Generate Mean Areal Precipitation

After reviewing the various methods available for generating spatially distributed rainfall, it was determined that the PRECIP program was the best approach. PRECIP is a program that was developed by the Corps of Engineers Hydrologic Engineering Center to generate Mean Areal Precipitation from observed data. PRECIP is capable of using irregular time steps, data sets with missing data and data beginning and ending at different times. An approach and steps for this project were outlined by David Ford Consulting Engineers staff and given to the LCRA staff.

The limit on number of gages used in the PRECIP program used was 400. This limit refers to the David Ford Consulting modified PRECIP program. The number of gages did not push the limits of the PRECIP program.

The limit on sub basins is 300 in the PRECIP program. This limit would have been a problem if the sub basins were not divided into and placed in two PRECIP files in the beginning. Later it was determined that more PRECIP files were needed. These two files were separated by grouping the upper basin together where the rainfall in general was less than 10 inches. Initial runs had results too high for the Pecan Bayou and upper sub basins because the defined maximum distance allowed too much influence of some of the intense centers. The final run grouped the sub basins to allow a better definition of which rainfall gages should be included in the calculation of the average Mean Areal Precipitation. The centroids of the sub basins were calculated using GIS and entered into the PRECIP program during the initial run. However, in the areas where there was a significant variation of rainfall, or in basins that were long and slender, more than one point was needed to represent the sub basin.

The maximum allowable distance to the gages and the sub basin points was initially entered as 30 miles except in the lower basins (below Mansfield dam) where the distance had to be expanded to 40 miles. The final run had four maximum allowable distances which were 10, 14, 24, and 30 miles. This was needed to better produce realistic results where the rainfall varied from 3 to 20 inches in a matter of 10 miles. To be able to have flexibility with areas of influence or the maximum allowable distance, multiple PRECIP files were set up to run via a batch file. Figure 1 shows the batch file commands and the necessary command lines to run the various files.

```
Precip I=1938storm_6HR.pre ads=precip1938 O=precip_6H.out
precip i=1938storm_6HRu.pre ds=precip1938 O=precip_6Hu.out
precip i=1938storm_6HRu10.pre ds=precip1938 O=precip_6Hu10.out
precip i=1938storm_6HRu15.pre ds=precip1938 O=precip_6Hu15.out
```

Figure 1. Batch file for running HEC-PRECIP Program

Results of PRECIP Program

Mean Areal Precipitation was produced for all the sub basins using at least one gage within 30 or 40 miles of the centriod of the sub basin with the exception of one sub basin. The subbasin was LC-18 which is the last basin on the Colorado River nearest to the coast. Since no rainfall gages were found that could supply the data needed to produce the Mean Areal Precipitation, the sub basin was assumed to have zero rainfall. This is a safe assumption for the 1938 flood because most of the rainfall was centered over the upper basin. An artificial gage with zero rainfall was placed in the LC-18 sub basin.

One of the outputs of the PRECIP program is the average distance between gages. The average distance between gages was about 10 miles with the sub basin Mean Areal Precipitation being computed with an average of 8 rainfall gages used for each sub basin. The range of the number of gages for the computation of Mean Areal Precipitation was between one (1) and twelve (12). Most of the lower basin below Mansfield Dam had only one gage that could be used. This area had little recorded data but only received small amounts of rainfall during the 1938 event.

The output of the PRECIP program contains a table that has a listing of the sub basin, a weighted distance, a list of the gages with the distance to the gages used by the program, along with percent of time the listed gage was used. This table was used to look at the results of the PRECIP program to determine if it was producing reasonable results and using the gages that had been entered.

Another tool that was used to look at the results of the Mean Areal Precipitation was displaying them in GIS. A shape file was created using the total Mean Areal Precipitation results from PRECIP and comparing it to the pseudo grid produced and the total of the gages themselves. In reviewing both the pseudo grid and comparing it to the USGS isohyetal map and the Mean Areal Precipitation, it seemed that PRECIP produced reasonable results. USGS isohyetal maps were also digitized showing two-day totals and these were used to look at the PRECIP results. Visualizations on a daily time-step were created using both rainfall data and the Mean Areal Precipitation. Each time step was reviewed to ensure that the results of the PRECIP runs were close to the observed results.

Results from Hydrologic HEC-HMS Model

HEC-HMS version 2.2.2 was used to run the hydrologic model. The model of the Colorado Basin is 21,745 mi^2. This includes the area of the Colorado River above O.H. Ivie Reservoir to E.V. Spence, Twin Buttes, and O.C. Fisher Reservoirs and extending downstream to Matagorda Bay. The PRECIP program results were input to the HEC-HMS model and loss rates and routing parameters were calibrated using historic USGS flow hydrographs. The observed runoff volume and peak discharges were compared at various sites. Table 2 is the summary table for the observed data versus the HMS results.

Location	Total Volume (ac-ft) - July 20-30, 1938		Peak Flow (cfs)	
	Observed	HMS Results	Observed	HMS Results
Colorado River at Ballinger	88,500	87,100	21,100	20,100
Concho River near Paint Rock	211,500	209,700	86,000	85,100
Pecan Bayou at Brownwood	26,000	26,000	5,300	3,800
San Saba River at San Saba	719,200	735,100	203,000	164,400
Colorado River near San Saba	1,928,000	1,958,000	224,000	330,200
Llano River near Castell	295,400	295,900	133,000	99,600

Table 2. Summary of Observed vs. Modeled Flows at Various Points in the Colorado Basin

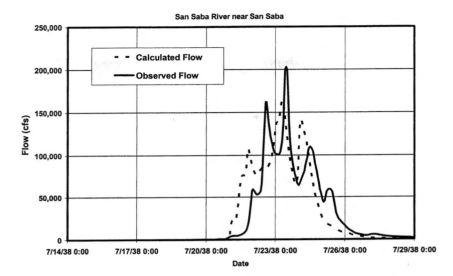

**Figure 2. -Comparison of Observed vs. Modeled Flows at San Saba River
near San Saba, Texas Gage**

Although runoff hydrograph volume totals matched fairly well between the observed values and the HEC-HMS generated values, the peak flows and timing was not exactly as it was observed. Comparisons of the observed versus model flows are shown for the San Saba River near San Saba, Texas in Figure 2. All the available gages were compared to see how well the computed matched hydrograph compared. There were several reasons for the differences in peak flow and timing of the runoff. The rainfall data was based on 6-hour intervals (longer intervals in some locations) and the HEC-HMS model was executed on a 15-minute time-step. This difference in time intervals led to differences in the peak flows and runoff hydrograph timing. Routing issues in the HEC-HMS model were another source of peak flow and timing differences. The routing data used in the HEC-HMS model was based on current channel geometry conditions. The channel geometries have most likely changed over seventy-years, especially with the increase in reservoir regulation within the watershed impacting erosion, sedimentation, and normal low flows within the reaches. Since most of the rainfall observations were from rainfall observers and not automated gages, some of the differences may be due to the fact that the data recorded was not what actually happened. It may be that the rain fell at different times but the observer could not get to the gage to check.

Future Applications and Suggestions

Although only the 1938 storm was re-created in this effort, the PRECIP file created can be used in other re-creations along with some of the methods for the display of GIS data.   Hourly data exists for the 1952, 1957 and 1978 (Pedernales River flood); storms, so these could also be re-created with more refined rainfall data than the 1938 work. The methodology applied in this study could be used to transpose storms over other basins. The method or approach can be used to create model data to better calibrate models where no existing calibration data is available.

Some of the suggestions or lessons learned in the project include:

- Know your version of PRECIP and its limitations. The limitations for the version used in this work were: 300 sub basin limit, 400 rainfall gages, miscellaneous and index gages, and a five point limit to describe the sub basins.
- Do one thorough research effort of available data at the onset of the project.
- Even though it is a lot of work to import data into GIS, it will help visualization and QA/QC. It is worth the effort. Do extensive QA/QC of the data, both input and output. This will give a better chance to produce good results.
- Find a stopping point in the project. It may be that not enough data can be found to generate reasonable results or any additional data you find will not improve the results for both volume of water and peak flow.
- Ten square miles seems to be a good rule of thumb for a radius of influence for a rain gage. If not enough gages can be found, a storm may not be reproducible due lack of information.

Conclusions

Overall, the re-creation of the July 1938 flood yielded a close runoff volume to observed hydrographs. In the operation of the reservoirs, volume and timing are especially critical. Re-creating the flood was successful in the fact that trends and volumes could be reproduced. The hydrologic model was calibrated to the volume of the storm and not the peak. The calculated peaks of the hydrographs were on average within 21% of observed values. This may seem high but there are many reasons for the differences. The reasons include:

- The majority of the data was in a 6-hour increment.
- Only thirteen USGS gages were available to do comparisons. This left large areas between calibration points. More observed data may have improved the peak flow results.
- Basin changes over 70 years may have resulted in different routing times.
- Considering wind and that the majority of the gages were not automatically reporting rainfall, the rainfall data may not be 100% accurate.

Considering the limited availability and date of the data used for this re-creation, we were pleased with the results. This project allowed engineers to see how to use observations from rainfall observers and look at the limitations of rainfall data.

Acknowledgements

Several people worked on this project. The project consisted of a significant amount of data input. Everybody that worked on this project helped in the success of creating the final run and their work is very much appreciated.

The initial data input was made by Terese Barnett, a LCRA summer intern. She placed the initial data into Excel spreadsheets. This included the daily rainfall for seventy gages and the hourly flow data for the stream gages. This made the data easier to view in a graphical format and also made it more accessible to a number of people. Richard Diaz of the LCRA Technology services staff used spreadsheets that contained daily and hourly rainfall information based on location and precipitation amounts to display the information in GIS. He also digitized the isohyetal maps found in the USGS reports to help QA/QC the data visually. Richard Diaz also used GIS to create the latitude and longitude information for the centriod of the sub basins which was a necessary input into the PRECIP program. He also assisted Dan Yates in the use of the model builder applications. Brad Moore, David Ford Consulting Engineers, outlined the approach to the project and he helped look at some of the initial error messages that PRECIP was giving. Brad Moore also listened to LCRA's questions on the PRECIP program and rainfall throughout the project. Rochelle B. Huff with David Ford Consulting staff helped by reading the PRECIP source code, providing us an idea of the limitations we had to work with, and helping us understand the PRECIP error messages.

Dan Yates helped input some of the 6- hour data along with importing the Mean Areal Precipitation results several times into GIS to look at the results of the PRECIP program. He also created the pseudo grid of the total rainfall using rainfall gages that were available for the study. This was eventually completed with the use a of model builder and resulted in the creation of the display of daily totals. He cycled through these tasks several times as each run often required a visual check. Melinda Luna did the initial research of the data available for the study, did the initial QA/QC of Terese Barnett's work, imported the Excel spreadsheets into DSS, created the PRECIP input file, entered 6- hour, hourly and daily data used in the final run, ran the PRECIP program, created the files that were used in the creation of some of the GIS displays, and compiled this report. Melinda Luna also coordinated the work with everyone involved and submitted the final rainfall products to Halff Associates. Halff Associates worked to get the rainfall input data into the hydrologic model. Halff created the 21,745 square mile HEC-HMS model and calibrated the loss rates and routing parameters to observed USGS flow records for the July 1938 storm event. This effort required several iterations with LCRA staff until a final rainfall data set and the calibrated July 1938 HEC-HMS model were developed.

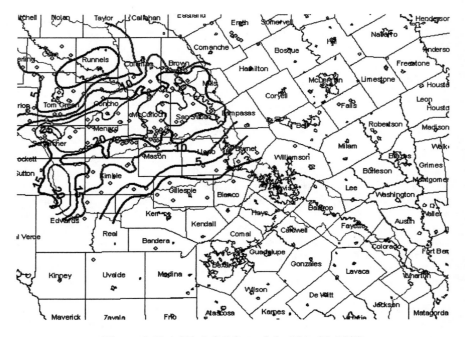

Figure 3–Total Rainfall from July 16 to 25, 1938

References

Breeding, S. & Dalyrmple, T. (1944) Water Supply Paper 914, Texas Floods of 1938 & 1939, USGS

Corps of Engineers, (July 17-25, 1938) Precipitation Data – Recording Gage Stations

Corps of Engineers Study of the storm of 17-25 July 1938(August 2, 1944), Transmitted to LCRA

PRECIP (updated by David Ford) Program User's Manual-29 July 1986 Version (July 31, 1987), US Army Corps of Engineers

Subject Index

Page number refers to the first page of paper

Author Index

Page number refers to the first page of paper